Information Generativity:

Volume I: The New Science of Information

Robert R. Carkhuff, Ph.D.

Christopher J. Carkhuff, D.Sci.

Published by: HRD Press, Inc.
                22 Amherst Road
                Amherst, MA 01002
                413-253-3488
                1-800-822-2801
                413-253-3490 (fax)
                www.hrdpress.com

ISBN 978-1-61014-391-2

Editorial services by Robert W. Carkhuff
Production services by Jean S. Miller
Graphic drawings by Rigoberto Cruz-Mejia
Cover design by Eileen Klockars
Promotion by Swift Global Media

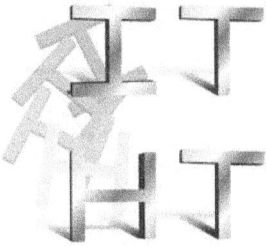

- This logo represents **generativity:** the interdependent (↔) and synergistic **processing** of information technology **(IT)** and human technology **(HT).**

$$\boxed{\text{IT} \leftrightarrow \text{HT}}$$

- This interdependent processing yields human capital development **(HCD)** and information capital development **(ICD)** that, together, account for 80% to 90% of the generativity within all organizations, communities, cultures, and economies:

$$\boxed{\text{HCD} \leftrightarrow \text{ICD}}$$

RRC
CarkhuffGenerativityLibrary.com

McLean, VA
January 1, 2015

Information Generativity:
Volume I. The New Science of Information

Contents

About the Authors

Among the most-cited scientists of the 20th century and already the most prolific in the 21st century, Robert R. Carkhuff is chairman of **The McLean Project** and the author of *The Human Sciences.*

His full body of work may be viewed on the websites:

www.mcleanproject.com

www.carkhuffgenerativitylibrary.com

Carkhuff boldly confronts our current socioeconomic crises:

"Generativity is the solution!"

"What is the question?"

Christopher J. Carkhuff is CEO of **GenStar, LLC,** a corporation dedicated to software development in the human and organizational generativity areas.

Preface

Information—Totally New Phenomena

The most fundamental assumption of science is this:

Science is a process.

Scientific products such as information may be found in learnable content. However, because the scientific products are changeable, they are not science per se. Function outputs may be processed as component inputs in continuous interdependent processing.

The second most fundamental assumption of science is this:

All science begins with the sensory experience of the scientist.

The scientist-observer is the prepotent source of scientific discovery.

This book presents a totally new way of discriminating and processing information. Assuming that information is the message being conveyed, the **New Science of Information** operationalizes, dimensionalizes, vectorializes, and phenomenalizes information by processing and "cubing" it into its vital dimensions. Thus, the product generated by processing is most powerful and actionable. Above all, it is of inestimable worth, increasing in quality with each processing iteration.

RRC

January 1, 2014
McLean, Virginia

Foreword (Part I)

The New Science of Information—
A Special Perspective

Dr. Robert Carkhuff has taken me on a **journey of discovery.** During this journey, I have been privileged to have experienced two special perspectives in my professional life:

1. The first was as one of the directors of IBM's Advanced Systems Development. There in 1968, I directed the **ECES Project: Educational and Career Exploration System** into a computer-based career guidance system. Stimulated by Carkhuff's Human Processing Systems, ECES evolved into the first-ever system empowered by *human centricity:* values-driven processing of job requirements. ECES became the new paradigm for educational and career processing in the 20th century.

2. The second of my special privileges was being introduced to Carkhuff's vision of the **New Science of Information.** Beginning with the most basic letters of the alphabet, Carkhuff has gone on to scale the levels of information:

   ### LEVELS

1	**Conceptual** or language-driven sentences
2	**Operational** or systems-driven operations
3	**Dimensional** or multidimensional schematics
4	**Vectorial** or socially-related schematics
5	**Phenomenal** or spatially-driven curvilinear schematics

Against this information scale, Carkhuff has set the **Human Processing Phases:**

   ### PHASES

1	**Relating by Goaling**
2	**Representing by Scaling**
3	**Reasoning by Exploring**
4	**Reasoning by Understanding**
5	**Reasoning by Acting**

The purpose is to process the appropriate levels of information in order to discover the most powerful constructs of information to provide the greatest perspective of human meaning. In other words, the goaling of information is to accomplish the mission of processing:

The "best fit" of human values with contextual requirements.

To be sure, Carkhuff has continued the evolution of information as well as human processing.

One is not possible without the other.

It is in this vein that Carkhuff has discovered the motherload of **Information Capital Development** or **ICD:** Both **Information Representation (IR)** and **Human Processing (HP)** relate synergistically to contribute to each other's growth.

$$ICD = (IR \leftrightarrow HP)$$

John T. Kelly, D.Sci.
Director Emeritus,
Advanced Systems Development, IBM, Inc.

January 1, 2009
McLean, Virginia

I am convinced that someone will eventually come up with a theory whose objects, connected by laws, are not probabilities...

— Albert Einstein
Letter to Born

I

The Universes of Information

1
The Possibilities Sciences

When the Judaic-Christian theologies gifted us with *"a knowable God,"* they provided the platform for science: *the explication of the unknown.* The more we could probe the unknown, the closer we could get to knowing God.

Thus, science is inherently a spiritual quest. It is a search for the *soul* of the universe. It is a search for the *souls* of the humans who occupy some small space within that universe.

Science is a search carried out by these most humble of God's creatures, seemingly made in His image. It is a search carried on by the greatest of God's gifts—human brainpower, with its infinite capacity to generate and organize itself to capture and fulfill God's universes.

Science is the search that addresses the *nature of nature.* Indeed, the nature of nature is precisely the same as the nature of the scientist-biographers who pursue its secrets.

The nature of nature is social. The functions of nature are to relate: all of its systems relate interdependently within, between, and among themselves.

The nature of nature is informational. The components of nature are informational: all of its systems communicate informationally within, between, and among themselves.

The nature of nature is processing—interdependent processing: all of its systems process interdependently within, between, and among themselves.

Indeed, all of nature's systems are processing systems. It is only when the scientist-biographers utilize brainpower systems to generate images of nature's systems that the scientists and nature become *one.* It is only when the scientist-biographers process interdependently within, between, and among nature's systems that we come close to knowing God.

Probabilities Science

The first of human efforts to formulate scientific systems to study phenomena may be termed *probabilities science.* Probabilities science enables us to *control* phenomena within a small *window* of opportunity in space and time. *Control* is the operative word here, for the functions of all probabilities science culminate in control: we describe phenomena in order to predict them; we predict phenomena in order to control them for humankind's purposes. These are the fundamental functions of probabilities science: describing, predicting, controlling.

The assumptions of probabilities science enable us to accomplish this *control.* The first of these is the linearity of phenomena. Scientists assume that, in light of human conditioning, there is no real difference between human processing and the linear processing of mechanical systems; indeed, for many, humans continue to be modeled upon linear mechanical systems (see Figure 1). Even the branching systems of information technologies are based upon *go—no go* choices of linear systems: the discriminative learning systems that enable choice empower us to choose only between one linear system and another. In other words, *truth* is a *line:* a scale or measurement or program to achieve an objective.

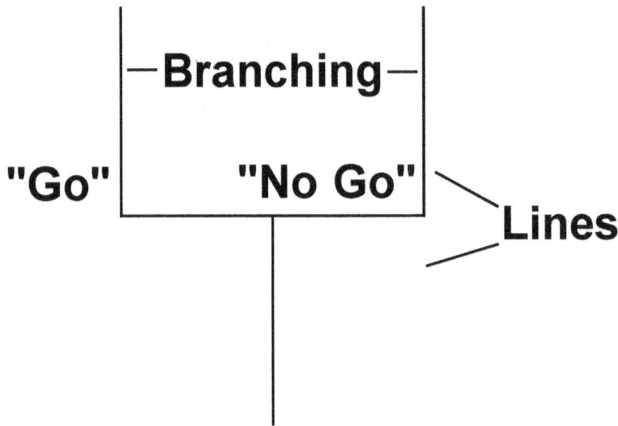

Figure 1. The Probabilities Assumption of Linearity

The second of these assumptions of probabilities science is the independence of phenomenal factors in the universe. Independence is assumed in orthogonal terms: one factor is as likely to relate to another factor as it is likely *not* to relate at all (see Figure 2). In other words, the relationship between independent factors is random.

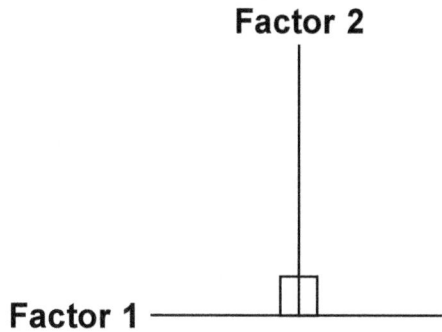

Factor 2

Factor 1

Figure 2. The Probabilities Assumption of Independence

The third of these probabilities assumptions is the symmetry of the phenomenal curves. Symmetry is assumed in terms of the *normal* distribution of relationships between and among independent factors; even skewed curves are treated as *normal* curves in terms of the analyses of their central tendencies and variabilities (see Figure 3).

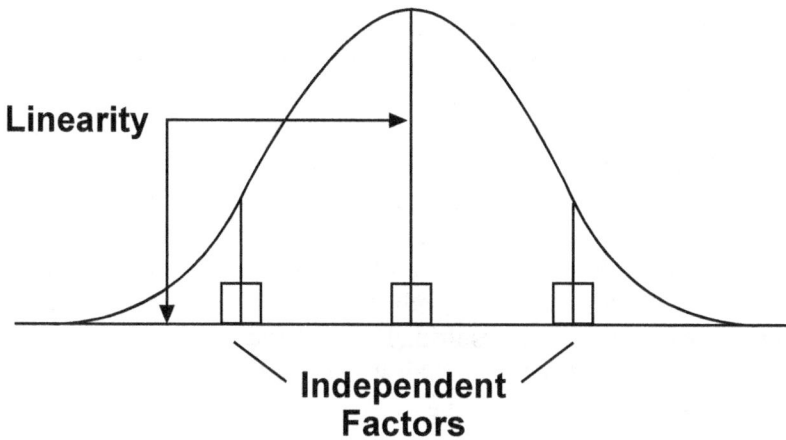

Linearity

Independent Factors

Figure 3. The Probabilities Assumption of Symmetry

The fourth assumption of probabilities science is the constancy or stability of these independent factors. Constancy is assumed in replicable terms: because the variability of factors is narrowed, the factors remain the same; even when we do secondary and tertiary analyses, we analyze the same data we used during our primary analysis, which yielded our original factors (see Figure 4). In other words, the phenomenal factors and their independent relationships are enduring; the phenomena are constant.

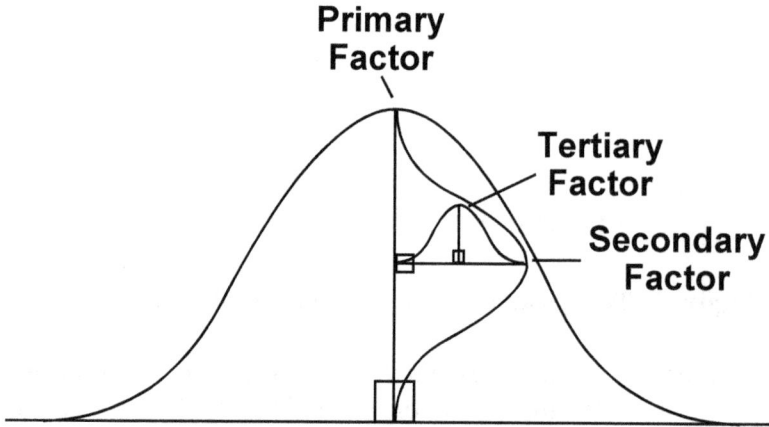

Figure 4. The Probabilities Assumption of Constancy

The fifth assumption of probabilities science is the source of constancy: content-centricity, the continuous emphasis upon specific content areas and levels. Content-centricity is the limiting principle of all probabilities science: the focus upon linear, independent, symmetrical, and static representations of specific content areas. In this context, not only the dimensions but also the content are exclusive and isolated and, therefore, doomed to attenuation by spiraling changes in the environment.

The implications of these assumptions of probabilities science are profound. The probabilities analyses result in local probabilities of static, symmetrical, and independent linear phenomenal factors. These phenomenal probabilities are due to artifactual data caused by the loss of variance resulting from artificial *controls*. We may term these phenomenal probabilities *node-centric:* their window of opportunity in space and time reflects only a data point in the expanding network of phenomena.

Possibilities Science

The next of human efforts to scientifically process phenomenal experience may be termed *possibilities science.* Possibilities science enables us to release or to free phenomena. The operative word here is *free:* the freedom of all phenomena, including human, is a function of the phenomena's processing systems; we relate to phenomena in order to comprehend their processing potential; we intervene to empower phenomena and thereby enhance their processing potential; we free phenomena in order to release their processing possibilities within God's universe of phenomenal possibilities.

The assumptions of possibilities science enable us to generate this freedom. The first of these assumptions is the multidimensionality of phenomena. Multidimensionality is assumed in the dimensions of the phenomenal processing systems: component inputs, transforming processes, function outputs, driving conditions, measurable standards (see Figure 5). In other words, phenomena, whatever their form, are inherently multidimensional.

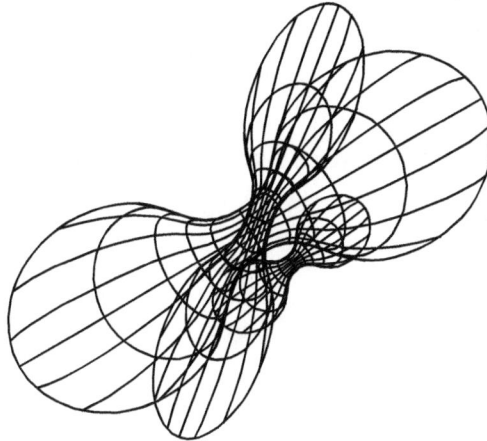

Figure 5. The Possibilities Assumption of Multidimensionality

The second of these possibilities assumptions is the interdependence of phenomenal vectors in the universe. Interdependence is assumed in vectored terms: this does not mean mutual dependency, but partnered processing for potentially mutual benefits; partnered processing between scientists and phenomena; partnered, virtual processing between phenomena and phenomena (see Figure 6). In other words, the relationships between interdependent vectors are interactive and synergistic.

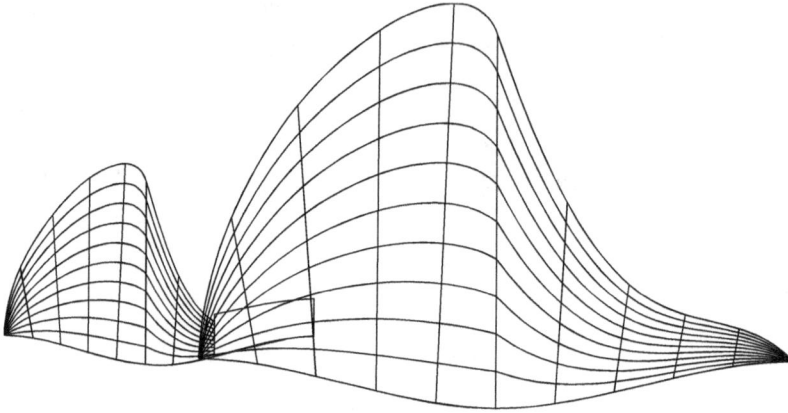

Figure 6. The Possibilities Assumption of Interdependence

The third of these assumptions of possibilities science is the asymmetry of the phenomenal curves. When we work with the data excluded from probabilities analyses (which assume independent factors and normal distributions), we find that asymmetrical curves define the essence of the phenomena (see Figure 7). We thus assume asymmetrical models of the changing nature of phenomena: there is no base line from which the phenomena vary, only infinite and asymmetrical changeability.

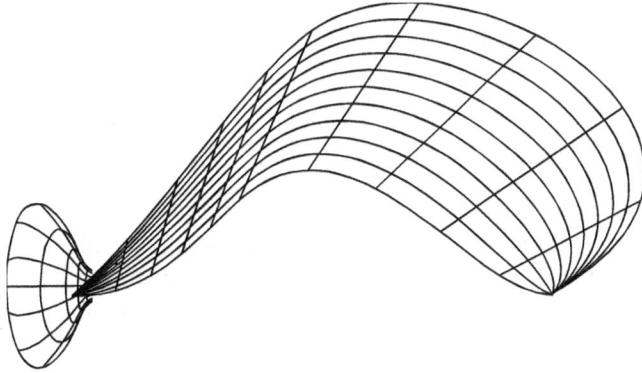

Figure 7. The Possibilities Assumption of Asymmetry

The fourth assumption of possibilities science is the dynamic and continuous changeability of these interdependent vectors. Changeability is assumed in terms of 360 degrees of global freedom or diversity: due to the continuous expansion of the changeability of *all* vectors, the changeability of *any* vector is continuous (see Figure 8). In other words, only the continuously changing phenomena are enduring: the only constancy is change.

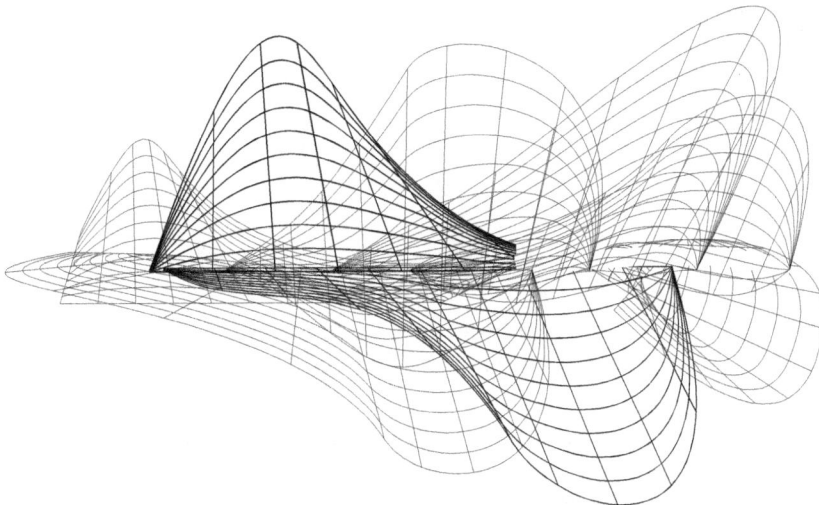

Figure 8. The Possibilities Assumption of Changeability

The fifth assumption of possibilities science is the source of changeability: *process-centricity,* the continuous processing of all phenomenal dimensions. Process-centricity is the generating principle of all possibilities science: the multidimensional, interdependent, asymmetrical, and changeable processing of all dimensions. In this context, not only the dimensions but also the processing systems themselves are continuously changing.

The implications of these assumptions of possibilities science are profound. Possibilities science generates expanding global possibilities of continuously changing, interdependent, and asymmetrical multidimensional phenomenal vectors. These phenomenal possibilities are due to the processing ability to align with the phenomena in their naturalistic form. We may regard these phenomenal possibilities as process-centric in a potentially infinite web of networks reflective of God's universes of changing phenomena.

We may view possibilities and probabilities phenomena in perspective in Figure 9. As can be seen, probabilities phenomena occupy a small window of opportunity in space and time. Indeed, we may think of them as *probabilities moments* rather than as phenomena in themselves. They occur within infinite possibilities phenomena. This is the essence of possibilities science: probabilities moments occurring within the context of infinite possibilities; possibilities phenomena ensuring that we have accurate phenomenal perspectives of probabilities.

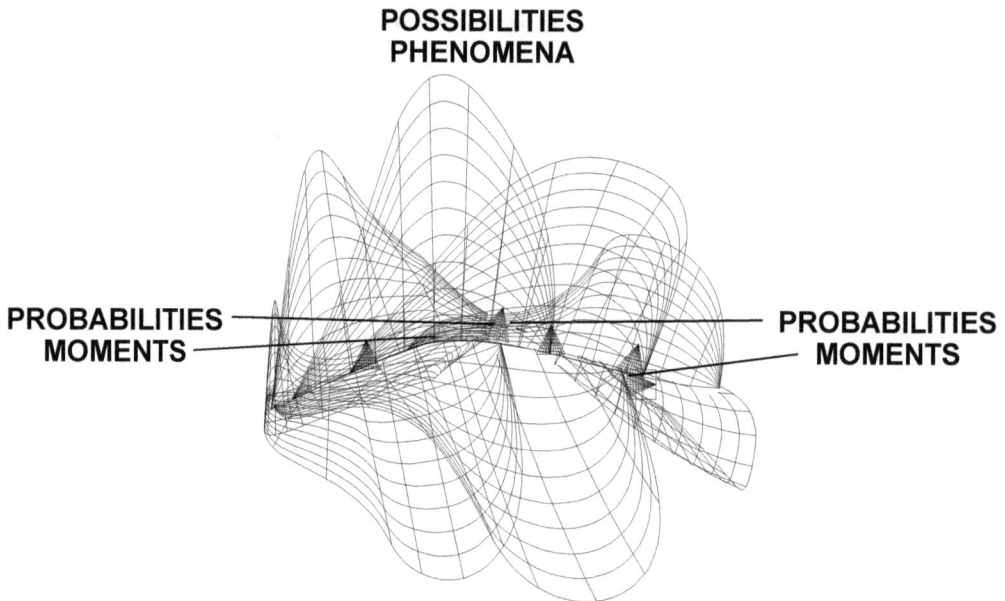

POSSIBILITIES PHENOMENA

PROBABILITIES MOMENTS

PROBABILITIES MOMENTS

**Figure 9. Probabilities Phenomena in a
Possibilities Phenomenal Context**

When we are producing products and services of any nature—whether they involve people, data, or things—we employ possibilities to drive probabilities: possibilities ensure that we are *doing the right thing.* We employ probabilities to produce our products: probabilities ensure that we *do things right.*

In turn, our possibilities design incorporates probabilities. Surely, we may employ probabilities, such as our knowledge of *best practices* in specific areas, to empower scientists to generate possibilities designs. In this instance, our probabilities empowerment enables our possibilities design.

To introduce possibilities models representing phenomenal possibilities, we may employ probabilities imagery: linearity, right angles, singular curvilinear tracks, and the like. However, the asymmetrical nature of phenomenal possibilities provides the framework for empowering probabilities moments: whatever the nature of phenomenal processing, the phenomena will reconfigure asymmetrically. *Thus, we can now freely employ free dynamic models to provide perspectives for static models:* whatever the nature of the phenomena, they will change continuously—birthing, growing, dying, rebirthing—to discover and rediscover their own changeable forms.

We also emphasize information models of all phenomena, doing so because we can understand them most readily. That is the nature of possibilities science. If information lives, then everything lives, for we can generate information models for everything. In this context, all science is applied science: *everything lives!*

While these sciences and the phenomena they generate are interdependent and synergistically related, the power of one dominates the contributions of the other: possibilities are expanded, infinite; probabilities are reduced, infinitesimal. Possibilities have been God's province. Probabilities have been, until now, humankind's province.

Although humans can be justly proud of their probabilities contributions, those contributions are infinitesimal in relation to our infinite phenomenal universes. The entire history of civilization, for example, has revolved around energy: discovering, extracting, refining finite resources of fossil fuels—coal and shale, oil and gas. Indeed, in the twentieth century, we fought two great world wars primarily over energy, killing more than 100 million people and destroying the lives of hundreds of millions of others.

Now, drawing from the generativity of the fictional *Star Trek*, we employ ion-driven, solar electric propulsion systems to thrust our spacecrafts into the heavens. Someday, burning fossil fuels will simply be a memory trace for civilization: a *probabilities moment.* What kind of a civilization will we create when we learn to align, enhance, and release the infinite phenomenal resources of God's multiple universes?

No, the describing, predicting, and controlling functions of probabilities science are not going to define and actualize our brave, new, and prosperous world; neither are the parametric assumptions, planning paradigms, and statistical process controls!

Yes, the relating, empowering, and freeing functions of possibilities science are going to generate and innovate our continuously growing and changing human and phenomenal destinies! Our unfettered, paradigmatic assumptions and process-centricity are going to release the power of the universes to literally and physically make everything out of nothing but our brainpower and our information models: our precious science of possibilities that drives our technologies of probabilities.

Clearly, in possibilities science, we serve a freeing function. The power of possibilities science is found in the interdependent processing of matured scientists with the phenomena they are addressing. The interdependent processors discern and apply the powerful forces guiding our universes: multidimensionality, interdependence, asymmetry, changeability. Both processor and phenomena mature!

Possibilities science is both source and force; content and processing; means and ends of all possibilities. Only our egos and their illusion of independence prevent us from understanding the interdependent nature of the universe.

We may represent these phenomenal possibilities symbolically by continuously changing, multidimensional, interdependently related, and asymmetrical curvilinear dimensions (see Figure 10). Indeed, just as phenomenal possibilities are changing, they are also expanding: quantitatively in the direction and force of their vectors; qualitatively in their increasingly inclusive power to create space and time. They are like an all-powerful changeable crystal, constantly exploding in curvilinear space with new and interdependently related, multidimensional systems: each yields the infinite possibilities of the changeable wholeness of our multiple universes.

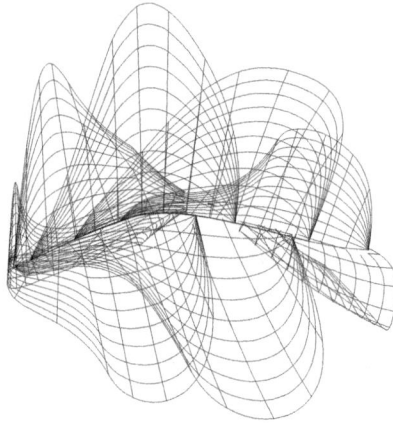

Figure 10. Generating Phenomenal Possibilities

As these universes of possibilities expand, they generate *probabilities moments:* opportunities upon which probabilities science may capitalize. As these universes of possibilities contract, they *become* probabilities phenomena. In this context, with the perspective of temporal and spatial distance, we may collapse this multidimensionality in singularity (see Figure 11). However, nature, having no tolerance for singularity, again explodes to generate new and changeable and expanding phenomena.

Figure 11. Singularity

II

The Experience of Information

2
The Phenomenal Experience

It is traditional for the philosopher to frame the following proposition:

If a tree falls in the forest and there is no one there, does it make a noise?

Now, with the advent of The Information Age, we may reformulate the question:

If a tree grows in the forest and there is no one there, does it communicate its existence in any way?

Think about it for a moment, for the answer has profound implications for the Science of Information!

The answer is **"Yes!"**

This green plant, which we label a tree, communicates its phenomenal existence in every way.

The source of this ubiquitous answer is information: **The representation of phenomenal experience** (see Figure 12).

As may be viewed:

- The genes empower the direction of growth through DNA.

- Photosynthesis directs the process of growth through the manufacture of simple sugar from CO_2 and water in the presence of light and chlorophyll.

- Oxygen, or O_2, is produced as a product that enables respiration as well as combustion necessary for life.

The Science of Information is represented at every phase of the green plant: input, process, output.

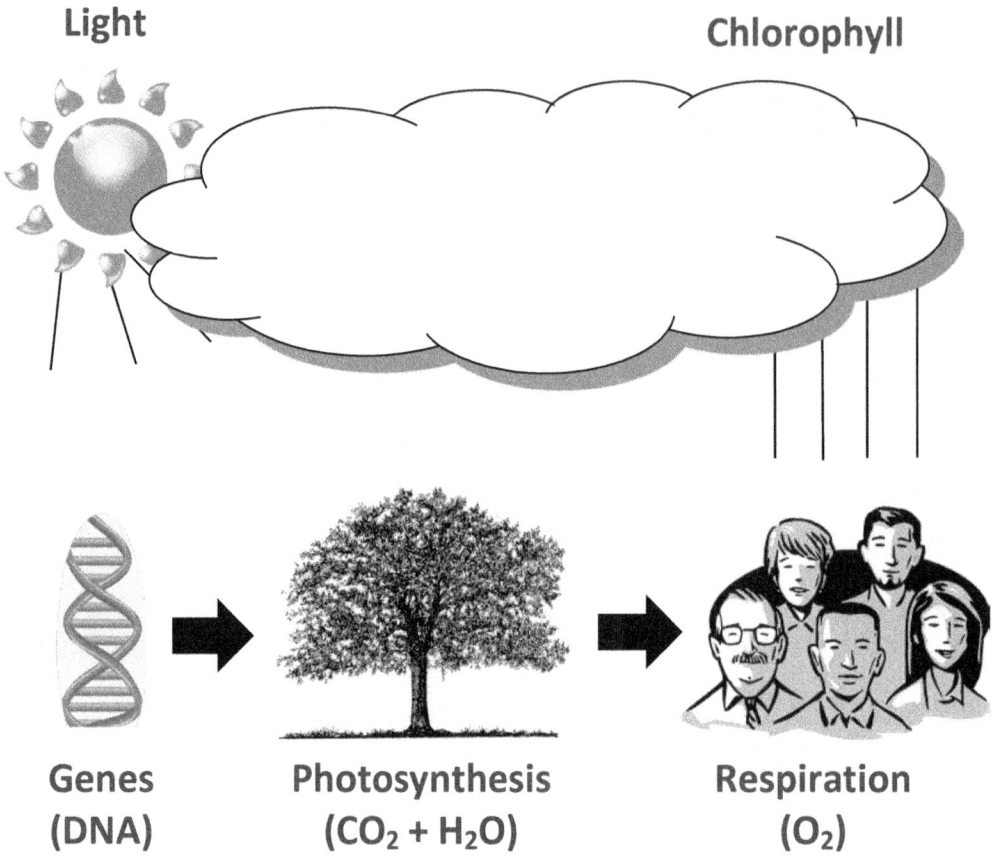

Light

Chlorophyll

Genes
(DNA)

Photosynthesis
(CO_2 + H_2O)

Respiration
(O_2)

Figure 12. The Information Representation of Photosynthesis

In a similar manner, we may view the phenomenal experience of the scientist (see Figure 13).

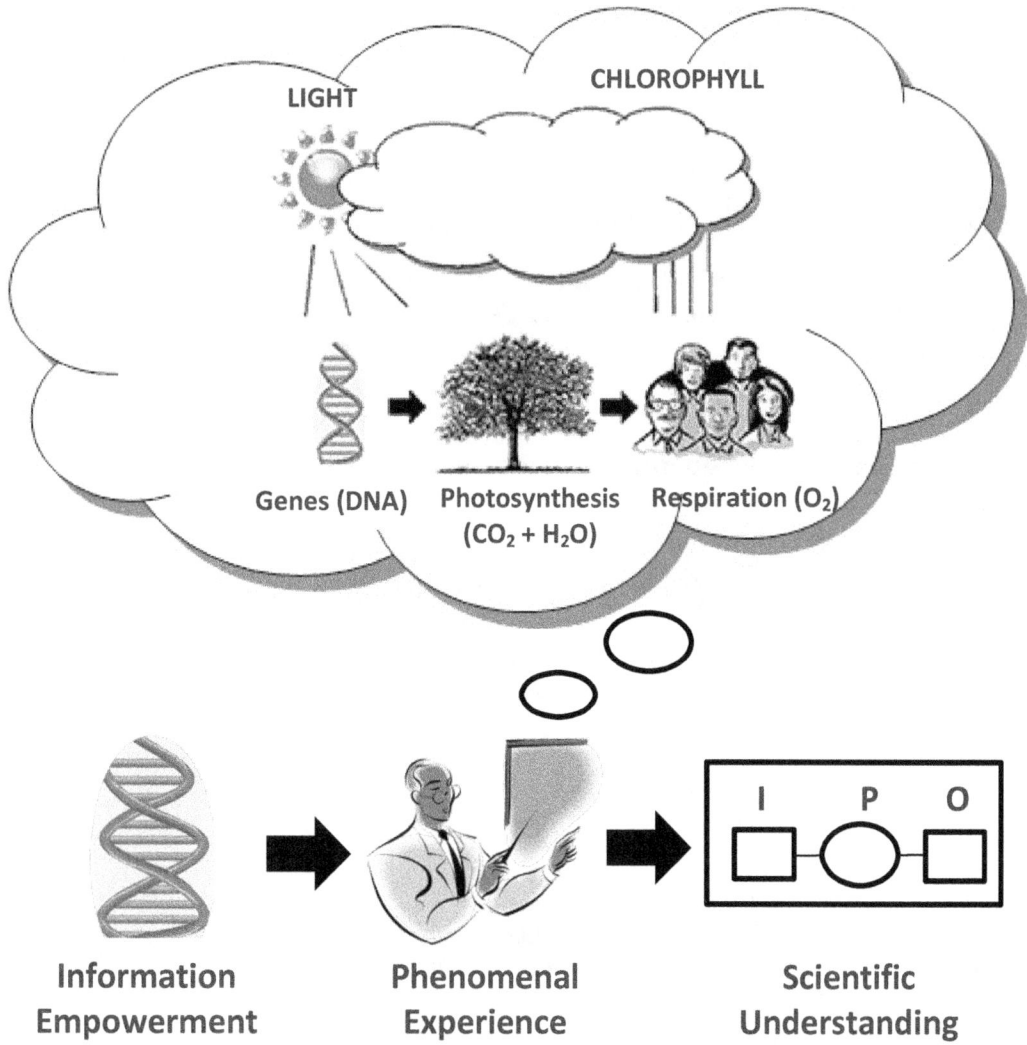

Information Empowerment   **Phenomenal Experience**   **Scientific Understanding**

Figure 13. The Phenomenal Experience of the Scientist

We may begin by asking the Information Origin question:

If science begins with the sensory experience of the scientist, where did the scientist's potential for phenomenal experience begin?

Again, the answer is DNA:

The programming of the response potential for representing phenomenal experience.

Likewise, we may ask the Information Understanding question:

If science culminates with the information modeling of the phenomenal experience, when is the scientific understanding complete?

Once again, the answer is the product produced information models.

The modeling of the operations of the phenomenal experience— inputs, processes, outputs

Information Modeling

Information modeling is the culmination of the phenomenal experience. Information modeling is complete when it meets all the criteria of science:

- Operational
- Replicable
- Measurable

In other words, the phenomenal experience is modeled in such a manner as to be transparent and sustainable. Two such information modeling programs are immediately available to all scientists: information systems and information cubes.

Information Systems

In Figure 14, we may view information systems that are complete because they define all the operations of any phenomenon:

- **Conditions** that influence environmental or contextual requirements
- **Functions** that define results outputs
- **Components** that source resource inputs
- **Processes** that define transforming programs
- **Standards** that define achievement feedback, which is recycled as input to the system

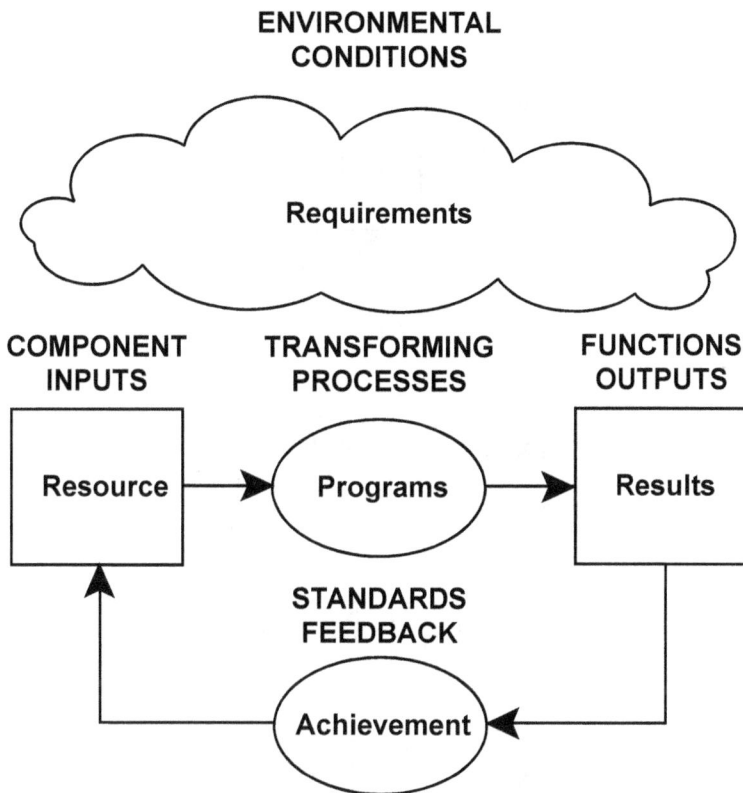

**ENVIRONMENTAL
CONDITIONS**

Requirements

COMPONENT INPUTS	TRANSFORMING PROCESSES	FUNCTIONS OUTPUTS
Resource	Programs	Results

**STANDARDS
FEEDBACK**

Achievement

Figure 14. Modeling Information Systems

Information Cubing

In Figure 15, we may view another mode of information modeling that is inclusive of all phenomenal operations. Labeled *information cubing,* this modality has the additional heuristic benefit of nesting multidimensional phenomenal requirements:

- *Nesting* or housing all lower-order phenomena in higher-order phenomena;

- *Encoding* or programming all lower-order phenomena by higher-order phenomena;

- *Rotating* or moving all lower-order phenomena to become higher-order phenomena and vice versa.

These principles empower the scientists to tailor information models that are developmental and predictive as well as functional and cumulative.

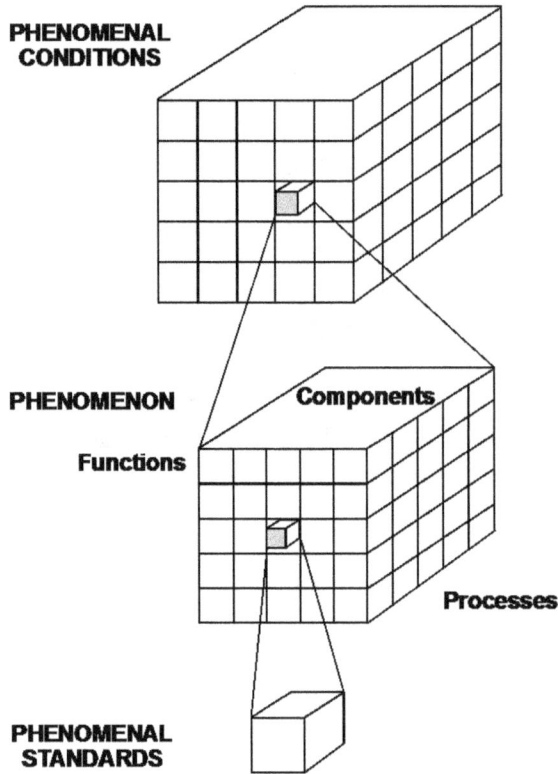

Figure 15. Modeling Information Cubing

Transition

We began by inquiring whether the Science of Information may be represented at every phase of phenomenal growth. The answer was a resounding **"Yes!"**

The phenomenal scientist may, himself or herself, be empowered to employ information modeling to define phenomenal growth and, in turn, to be modeled by it.

For example, as a scientist, when I am asked to express my objectives in phenomenal modeling, I begin with the operational definition illustrated in Figure 16:

- Phenomenal functions

- Informational components

- Human-information processes

In other words, the phenomenal operations may be expressed as follows:

Phenomenal functions are achieved by information components empowered by human-information processing.

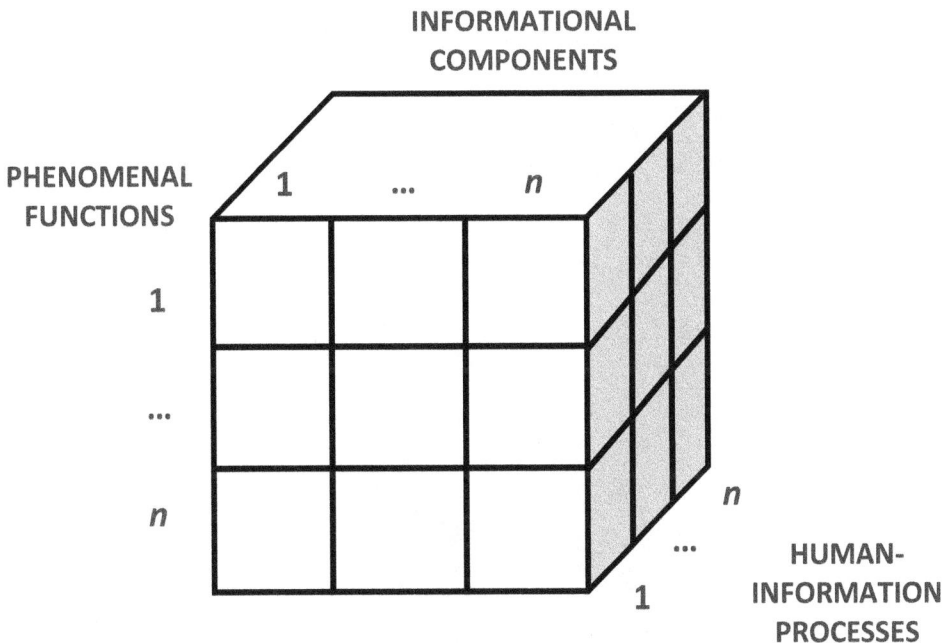

INFORMATIONAL
COMPONENTS

PHENOMENAL
FUNCTIONS

Figure 16. Operational Definition of the Phenomenal Objective

This is the basic objective in studying all phenomena:

- We embrace the functions of the phenomena.
- We express the levels of information components calculated to achieve the phenomenal functions.
- We empower the information components by interdependent and synergistic human-information processing.

It remains only for us to scale the phenomenal dimensions.

In transition, we propose the following:

The Science of Information is precisely this: The Information of Science.

The most heuristic of the information modeling modes is information cubing about which we will present in the pages that follow.

3
Information Science—An Introduction

Information! Nations treasure it. The marketplace requires it. Corporations generate it. Managers manage it. Yet no one seems to be able to define it clearly. In business, we speak about information by using terms such as *knowledge management,* but aside from our inventory and transaction databases, most of our files are filled with lots of reports and memos—often of little or no value. In fact, this wordy excess is a "depressor variable" for our productivity. Relatedly, while we promote the rapidity of our information *connectivity* capabilities, we still have little idea of how to develop and model valuable information to communicate across our networks. So what is information? How can we define and so develop and manage this information that everyone acknowledges is critical for economic growth?

To orient ourselves to defining information, let us begin by defining a few of the words that are used to describe it:

- *Information* is derived from the verb *to inform.* It implies modeling knowledge of the operational nature of phenomena. Information communicates operations: what something is, how it works, what it does, when and where it does what it does, and why it does it.

- *Knowledge,* in turn, is derived from the verb *to know.* It implies a conceptual understanding of the nature of phenomena and their relationships.

- *Modeling* is derived from the verb *to model.* It implies imaging, representing, or displaying the nature of phenomena.

Now let us put these definitions together: **Information emphasizes Representing: modeling to display images of the nature of phenomena.**

This book will introduce us to what we call *information science.* Information science will empower us to develop and model information at the highest levels of Ideation.

Conceptual Information

Perhaps the most ridiculous claims are made by those who know the least. Information technologists who have conquered the Binary Code of Processing have claimed ownership of the information science.

In turn, they have sought out the teachers at all levels of our educational systems to inquire of the science. The teachers have responded in kind:

> **"Information represents an understanding of facts, concepts, and principles of communication."**

In the teachers' minds, information represents the concepts, relationships, and objectives of communication. This is the first and most basic level of representation.

Labeled *conceptual information,* this level represents the verbal stimulation or comprehension of communication.

Usually conceptual information is recorded in verbal exchanges or written reports (see Figure 17). More than 99% of the reports filed in business record conceptual information.

Figure 17. Conceptual Information

Operational Information

What conceptual information lacks, operational information provides: the ingredients of action necessary for building or problem-solving. What distinguishes an engineer from a teacher, for example, is the actionability of their communications.

Operational information defines all phenomena in terms of the operations that they perform:

- **Functions** or results to be performed

- **Components** or resources to be invested

- **Processes** or procedures that transform resources into results

- **Conditions** or contexts that influence operations to be performed

- **Standards** or levels of achievement of operations

Together, the operations define the comprehensive system to be performed (see Figure 18).

CONDITIONS

Influencing Contexts

INPUTS	PROCESSES	OUTPUTS
Resources Invested	Transforming Procedures	Results Attained

STANDARDS

Levels of Achievement

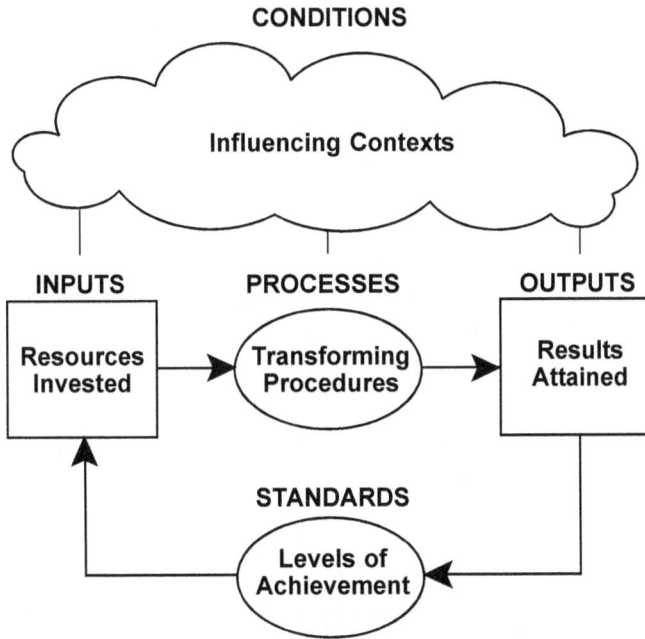

Figure 18. Operational Information

Dimensional Information

Where operational information defines the systems of phenomenal performance, dimensional information defines the performance in relation to the totality of its production. Where the engineer employs systems to achieve objectives, the architect "factor analyzes" schematics to relate all objectives.

In other words, each factored dimension is scaled in levels (1 – 5) and each level interacts with every other level (see Figure 19). In this manner, we introduce the interactivity that will extend into interdependency. This will enable us to elevate the objectives and increases the productivity of the systems.

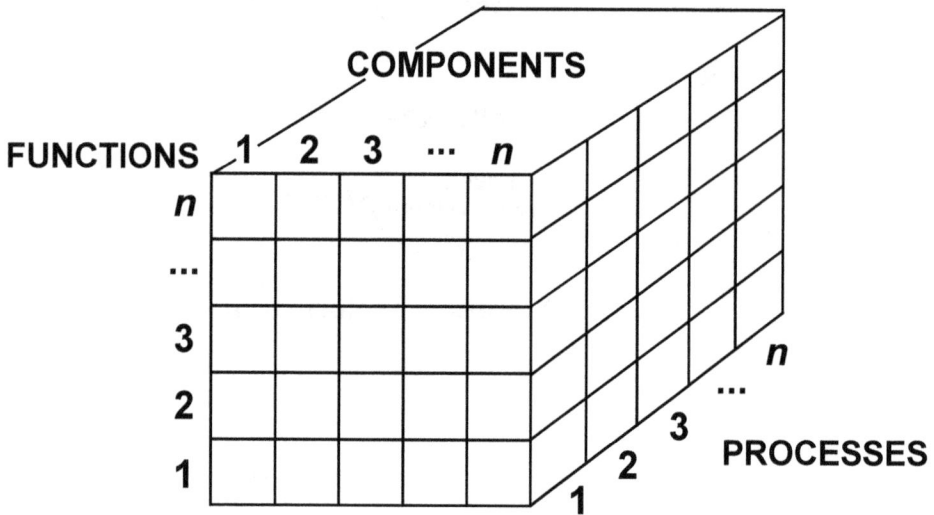

Figure 19. Dimensional Information

Vectorial Information

Where dimensional information factors and scales the defined dimensions, vectorial information factors and scales all dimensions impinging upon the phenomena. Where the architect designs the schematics to relate all dimensions, the technologist designs the social schematics of all related dimensions.

In other words, (1) all conditions from which the phenomena is sourced as well as (2) all standards to which the phenomena are dedicated are represented in their interdependency. As may be noted, all phenomena exist in interdependency with (1) the conditions from which they were drawn as well as (2) the performance standards to which they are dedicated (see Figure 20).

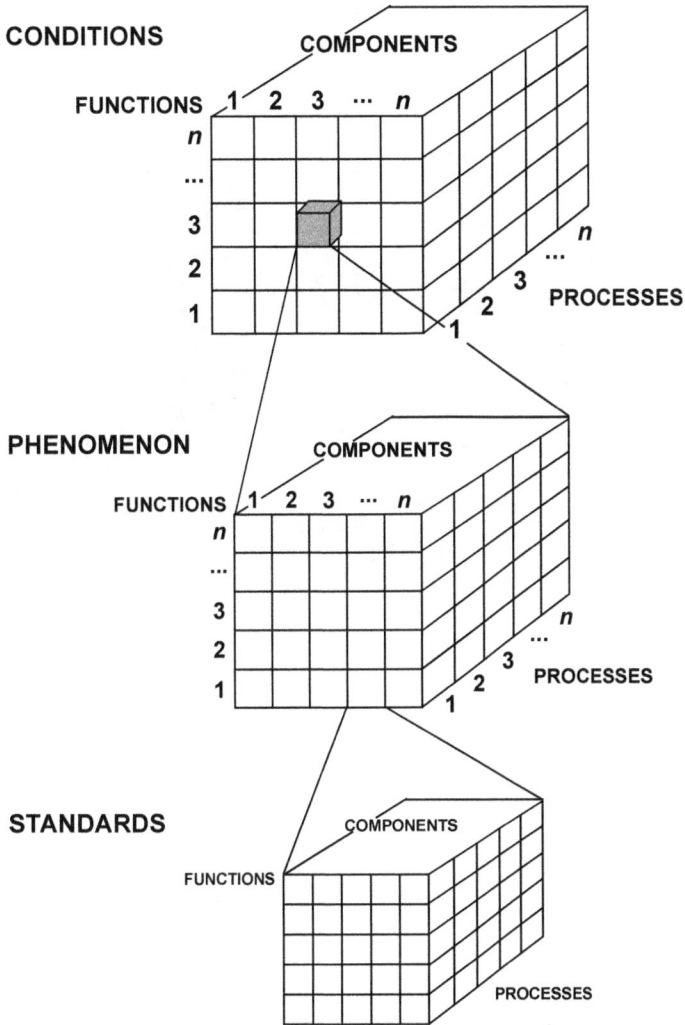

Figure 20. Vectorial Information

Phenomenal Information

Phenomenal information is where the scientific study of the nature of nature dominates. The scientist attempts to align (1) as closely as possible with the data on the nature of the phenomena, and (2) still make the data available to calculate and project the future courses of the phenomena.

In other words, scientists attempt to dimensionalize and vectorialize phenomenal information in which operations are inherently complex (see Figure 21).

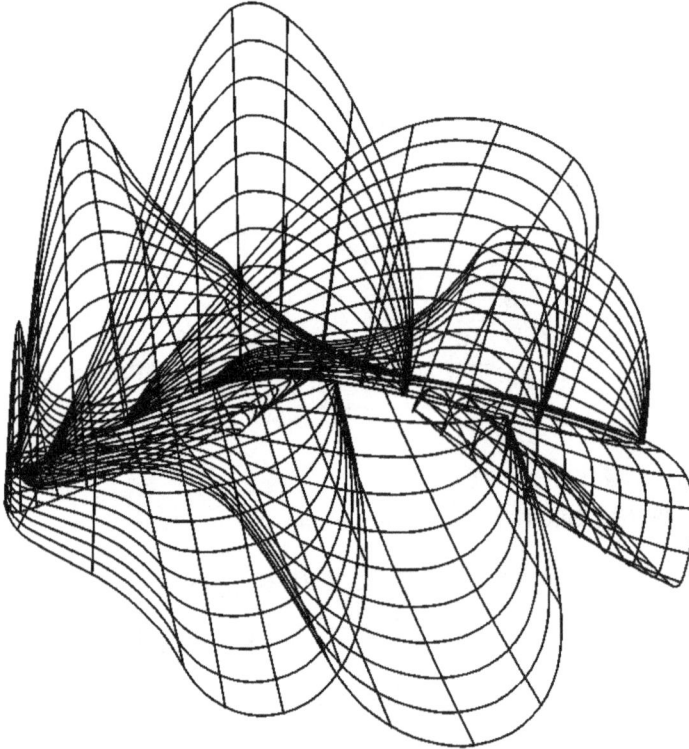

Figure 21. Phenomenal Information

Information Representation

Information representation is preparation for building information science.

Understand that phenomenal information extends the boundaries of interactivity to interdependency.

For the time being, suffice it to suggest that the defining and building of information science is the most complex of all human endeavors because of the inherent nature of all phenomena:

- Unequal in operations

- Multidimensional in components

- Interdependent in functions

- Curvilinear and asymmetrical in conditions

- Continuously changeable in standards

It will be our privilege to introduce the profound nature of the New Science of Information.

III

The Science of Information

4
Information Representing

We may build a model of *information capital* by representing the functions or outputs of information; its components or inputs; and its transforming processes—how these information components accomplish these functions. The functions are defined by what the information is intended to accomplish. The components of information are defined by their ingredients and how these ingredients are related. The processes of information are defined as the methods that can be applied to transform one form of information into another, and so serve the intended purposes of ICD.

In representing information, we define dimensions or ingredients of ICD as follows (see Table 1):

- **Conceptual** or verbal information

- **Operational** or systems information

- **Dimensional** or schematic information

- **Vectorial** or directional information

- **Phenomenal** or spatial information

Table 1. Levels of ICD

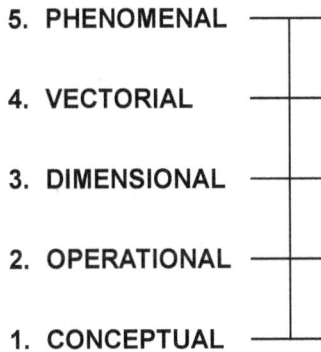

5. PHENOMENAL

4. VECTORIAL

3. DIMENSIONAL

2. OPERATIONAL

1. CONCEPTUAL

This scale of ICD is a useful initial *map-in* for analyzing the quality of our information. Further analyses of each of these types of information will provide us with a more detailed description of the kinds of information we need to develop.

By defining the components of information, we have tools for analyzing the quality of information. Do our information representations best serve their intentions? If not, then we should consider transforming those representations into other forms of information.

Operations and Applications of ICD

Before we can engage in meaningful information representing and productive knowledge management, we must expand our understanding about information beyond vague generalities. This involves defining the components of information capital to include the following:

- **Conceptual** information for representing relationships between phenomena
- **Operational** information for representing the operations of phenomena
- **Dimensional** information for representing the dimensions of phenomena
- **Vectorial** information for representing the directions and forces of phenomena
- **Phenomenal** information for representing the natural sources of the phenomena

By building these types of information, we can represent and manage our information modeling responsibilities.

Conceptual Information

By definition, *conceptual information* states *the relationship between things*. The value of conceptual knowledge is to state the relationships within, between, and among words or numbers or other schematic representations.

The levels of conceptual information are illustrated in Table 2. These levels have simple verbal definitions:

- **Facts** *identify labels* we attach to phenomena.
- **Concepts** *state relationships* within or between phenomena.
- **Principles** *explain relationships* within or between phenomena.
- **Applications** *specify contexts* in which the phenomena are demonstrated.
- **Objectives** *specify standards* for the performance of these phenomena.

Table 2. Levels of Conceptual Information

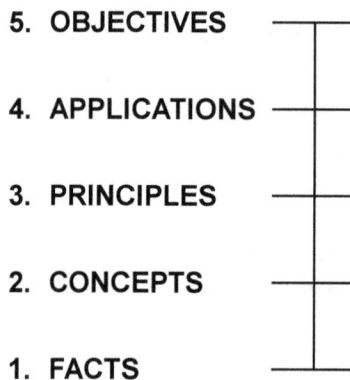

5. OBJECTIVES

4. APPLICATIONS

3. PRINCIPLES

2. CONCEPTS

1. FACTS

Conceptual information is the simplest form of information representing. We use conceptual information when we define phenomena *verbally:* things (facts) have relationships (concepts) that are explained (principles) in contexts (applications) so that we can accomplish standards (objectives).

Most of our information is represented conceptually. For example, facts and concepts fill our e-mails. When this information is further developed, we may write or read sentences that explain how certain facts are related. We may describe information about applications, or contexts, to tell when and where and perhaps why something is so. We may write or read information about standards of performance and why these standards are so.

Again, conceptual information is verbal — it is a string of sentences. It may be informative and useful, but it is also difficult to comprehend and remember. Accurate, comprehensive reports may be useful, but they will require a lot of effort to write and a lot of effort to read.

Operational Information

Operational information defines phenomena by their operations as parts of systems. We say that we have defined our information operationally when we have defined the phenomenal systems operations. In comparing this type of information to conceptual information, we find that conceptual information *describes relationships,* whereas operational information *defines the operations involved in the relationships.*

The levels of operational information are defined by their operational parts, as shown below (see Table 3). Operational information can be collected or generated by answering these basic questions:

- **Functions — What** are we trying to do?
- **Components — Who** or **what** is involved?
- **Processes — Why** and **how** are we doing it?
- **Conditions — When** and **where** are we doing it?
- **Standards — How well** are we doing it?

These are the ingredients of a basic unit of information.

Table 3. Levels of Operational Information

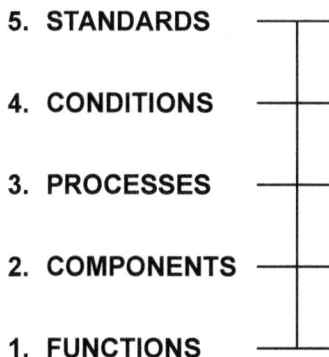

5. STANDARDS

4. CONDITIONS

3. PROCESSES

2. COMPONENTS

1. FUNCTIONS

In practice, operational information may be applied in systems, as shown in Figure 22:

- **Functions** — the products or services intended as outputs
- **Components** — the parts or participants invested as inputs
- **Processes** — the procedures or methods employed for transformation
- **Conditions** — the contexts or environments of the operations
- **Standards**—the measures of excellence or achievement for the operations

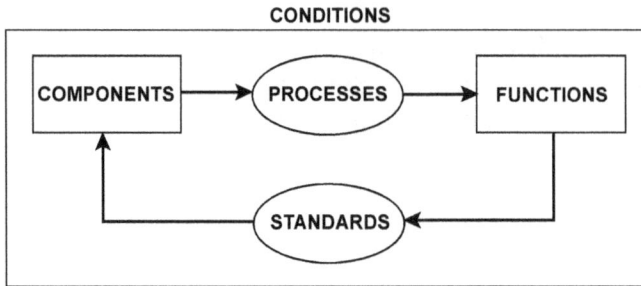

Figure 22. Operational Systems Information

When we understand that conceptual information can be defined in operational terms, we are able to transform and redefine facts, concepts, principles, applications, and objectives by their operations or systems parts. A matrix that communicates how conceptual and operational information is related is presented in Table 4.

Table 4. Operational Definitions of Knowledge

Levels of Conceptual Information	Operational Knowledge Systems
5 – Objectives	Conditions Components → Processes → Functions Standards ←
4 – Applications	Conditions Components → Processes → Functions
3 – Principles	Components → Processes → Functions
2 – Concepts	Components → Processes
1 – Facts	Components, Processes

As can be seen, by redefining conceptual information with operational systems, we have now defined the conceptual information as developmental and cumulative.

Conceptual information, then, is translated into operational information as follows:

- **Facts** are conceptually represented and generally understood as the labels we attach to things. Operationally defined, facts are the specific components and functions involved. Facts tell what we have done or what we are trying to do (functions) and who or what is involved (components).

- **Concepts** describe relationships within or between phenomena. Operationally defined, concepts are the relationships within and between components and/or functions; between components and components; between functions and functions; between components and functions. For ICD purposes, concepts describe what parts and participants (components) are related to what intentions (functions).

- **Principles** are explanations of the relationships within or between phenomena. Operationally defined, principles systematically explain the relationship of components and functions as well as the intervening processes involved. Principles include the steps (processes) that the parts and participants (components) will take to accomplish their intentions (functions). Principles tell us why and how the relationships transforming components into functions take place.

- **Applications** include information about the contexts in which phenomena are demonstrated. Operationally defined, applications are described as systems of components, processes, and functions, along with conditions or systems boundaries. Applications information tells when and where (conditions) the principles will be applied.

- **Objectives** become operational statements of standards or measures of performance. Operationally defined, objectives are described as systems of components, processes, functions, and conditions, along with information about measures of performance. Objectives tell how well (standards) the applications need to be performed.

Let us consider the differences between conceptual and operational information. Communication with conceptual information is the medium of conversation. It is also the medium of poets and novelists. We hear and read many interesting pieces of conceptual information, but it is left to our brainpower to put the pieces together. We hope to finally have an "Ah, ha!" experience. Conceptual conversations and reading can be highly rewarding. They are critical to relationship-building and may stimulate our thinking, however inefficiently.

In business, with customers paying for our time, we must be both effective and efficient in our communication. If we are not, our customers will become our former customers. By building information that is operational, by its systems parts, we cut through all those words. Basically, we discipline ourselves to develop our information in terms of systems: functions or outputs; components or inputs; processes or procedures; conditions or contexts; and standards or measures of performance.

Remember, conceptual information is *verbal*—it is a string of sentences. It answers the basic interrogatives. It requires a lot of effort to write and to read.

Operational information is *systems information.* We may represent it in a systems drawing or describe it in a series of sentences that focus on the operations of a system or systems.

Dimensional Information

What is *dimensional information?* From an operational systems perspective, we may say that each operation (system part) is a dimension of information. From this perspective, we see that there are five dimensions to any system: functions, components, processes, conditions, and standards.

Dimensional information presents this operational systems information as lists or scales; matrices or tables; models; *nested* models; and multidimensional *nested* models. The levels of dimensional information are defined by the complexity and nature of their dimensionality:

- **1D** — **One-dimensional** list or scale
- **2D** — **Two-dimensional** matrix or system
- **3D** — **Three-dimensional** model
- **ND** — *Nested*-**dimensional** models
- **MD** — **Multidimensional** *nested* models

We are all familiar with one-dimensional lists. Many of us know about some forms of scaling. Some of us understand two-dimensional matrices or tables. A few of us have been introduced to three-dimensional, *x-y-z* axis modeling. *Nested*-dimensional and multidimensional modeling display new axes or planes beyond three dimensions (see Figure 23).

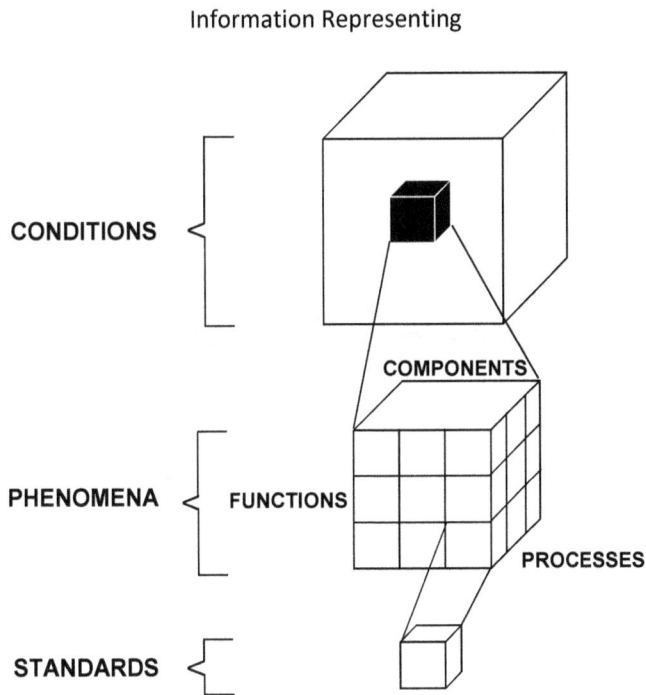

Figure 23. Dimensional Information

We may continue to represent these dimensions with additional three-dimensional models. As shown, the 3D model is *nested* in a higher-order model representing conditions. In turn, the 3D model *nests* lower-order models representing standards. All models of higher- and lower-order phenomena have their own dimensionality: functions, components, processes, conditions, standards.

Dimensional information representations are different from operational systems drawings. The lowest levels of dimensional representation, 1D lists or scales and 2D matrices or tables, are used to communicate partial operational systems information. The higher levels of dimensional representations can communicate the interaction of multiple systems. This information provides us with perspectives that are best represented in *nested*-dimensional and multidimensional modeling.

By building dimensional information, we discipline ourselves to develop our information with its multidimensionality in mind. Multidimensional modeling is a tool for communicating invaluable perspective. Wouldn't we be pleased if people who worked with us developed and delivered dimensional information to us? Then we could expand and narrow the products of our co-developed ideation by using the tools of dimensional modeling: 1D, 2D, 3D, ND, and MD.

Vectorial Information

Vectors are the forces of phenomena themselves. They exist, whether we recognize them or not. In nature, if a tidal wave is approaching land, it has force and direction, whether we realize it or not. Likewise, in the business environment, if a competitor's breakthrough product is about to be launched and will make our product obsolete, it is a vector about to smash us, whether we realize its vectorial power or not.

Vectorial information emphasizes facsimile representations of the actual forces of phenomena upon other phenomena. Vectorial information represents both the *direction* and *magnitude* of these forces. The direction indicates a particular course. The magnitude indicates the power of the force. The direction and magnitude of vectors may be represented by the convergence of forces, as illustrated in Figure 24.

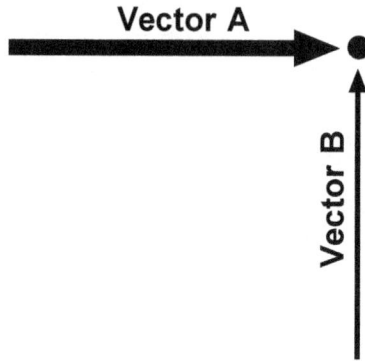

Vector A

Vector B

Figure 24. Vectorial Forces

In this instance, a phenomenon is about to be impacted by two vectors or forces. Both of these vectors will soon contribute to defining the context within which the phenomenon operates. As may be noted, Vector A is represented as having considerably greater force than Vector B.

We may also view an example of the effects of the direction and magnitude of vectors upon phenomena (see Figure 25). In this instance, the same two vectors are represented as having impacted upon a phenomenon at a certain point. The interaction of these vectors generates a resultant value.

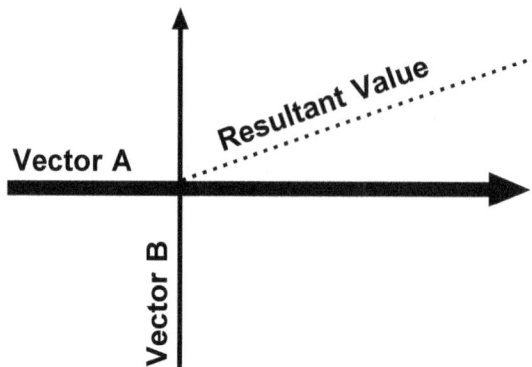

Resultant Value

Vector A

Vector B

Figure 25. Vectorial Forces and Resultant Value

We live in a world of vectorial forces and resultant values. From a defensive position, we hope that we can see the vectors coming, represent them informationally, and process their impact. From an offensive position, if we can see the vectors and represent them informationally before our competitors do, then opportunities are ours for the taking. For example, from a business perspective, the issues are straightforward: Do we recognize the vectors that impact upon us? Do we represent these vectors informationally so that we can impact upon them with intentionality?

In our own work, for example, we have researched the vectors of marketplace positioning, organizational alignment, human processing, information modeling, and mechanical tooling (see Figure 26). We define these forces in 3D models that we call New Capital Development or NCD models: MCD, OCD, HCD, ICD, mCD. We represent the force and direction of these vectors by the way we have nested them and by arrows that communicate deductive and inductive process movements. Once again, as we will discover, we use the basic information units to define all NCD operations: functions, components, processes.

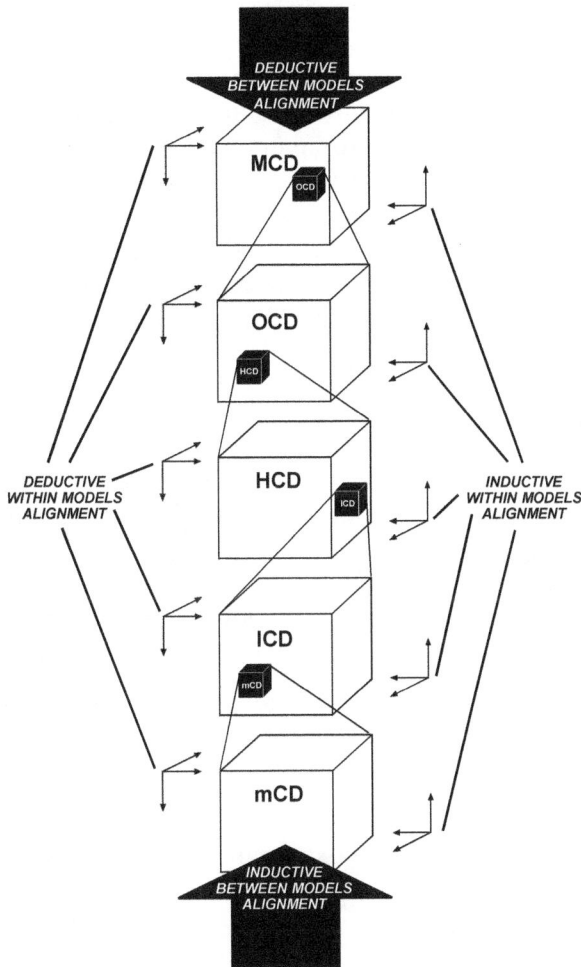

Figure 26. Vectorial Perspective of NCD Models

As may be noted, the levels of vectorial force and direction are defined by their NCD models:

MCD — **Marketplace Capital Development** or marketplace forces

OCD — **Organizational Capital Development** or organizational forces

HCD — **Human Capital Development** or human forces

ICD — **Information Capital Development** or information forces

mCD — **Mechanical Capital Development** or mechanical forces

Certainly, there are many vectors at work in our universe. We present our NCD systems as an example of our understanding of the powerful vectors at work in our human environments. We hope that these vectors prove to be valuable tools for generating wealth in the 21st century and beyond.

Vectorial information involves representations of phenomenal conditions. It is humankind's way of representing the force and direction of nature's phenomena. When we discipline ourselves to seek out vectorial information, we are determined in our search to understand the forces and the direction of forces acting upon phenomena. Without this focus, our conceptual, operational, and dimensional information may be inaccurate or irrelevant. When we represent vectors in our information representations, we communicate the contextual forces that are critical to us. With an understanding of vectorial information, we may intervene to enhance phenomena for humankind's purposes.

Phenomenal Information

Phenomenal information is the source of all information. Scientific relating to phenomena reveals the following shared characteristics of *all* phenomena:

- Phenomena are inherently **unequal:** no operations of any phenomena are equal.

- Phenomena are inherently **multidimensional and curvilinear:** a multitude of curvilinear dimensions interact to define the phenomena.

- Phenomena are inherently **interdependent:** the social aspects of nature assert that (1) nothing can exist by itself, (2) nothing can grow by itself, and (3) nothing can live or die by itself.

- Phenomena are inherently **changeable:** the changeability of nature itself asserts that all phenomena are continuously changing in qualitative substance as well as in quantitative form. Phenomena are always moving—birthing, growing, relating, dying, being born again.

Each characteristic is a required condition for the other characteristics.

Any generic representation of phenomenal conditions is captured at a *window of convergence* occurring at a particular point in space and time, as represented in Figure 27. As may be noted, multidimensional, interdependent, and changeable phenomena are represented here in a curvilinear form.

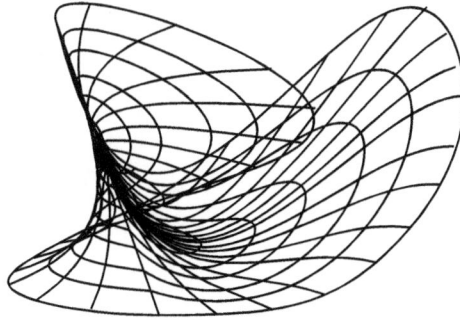

Figure 27. Phenomenal Information

A good illustration of phenomenal information in the business arena is the formation of a *market wave.* The wave may be generated by breakthrough technologies; however, it is formed in the commercial marketplace by the multitude of buying and selling decisions that occur each day in the marketplace.

Another example of phenomenal information is presented in Figure 28. In this instance, we researched data on the marketplace of mechanical and information technologies. Composed of literally trillions of data points, these phenomena are represented in curvilinear form to help discriminate and communicate them. As illustrated, the **Mechanical Technologies** or **mT** peak as a source of comparative advantage around 1960, while the **Information Technologies** or **IT** peaked in the 1980s.

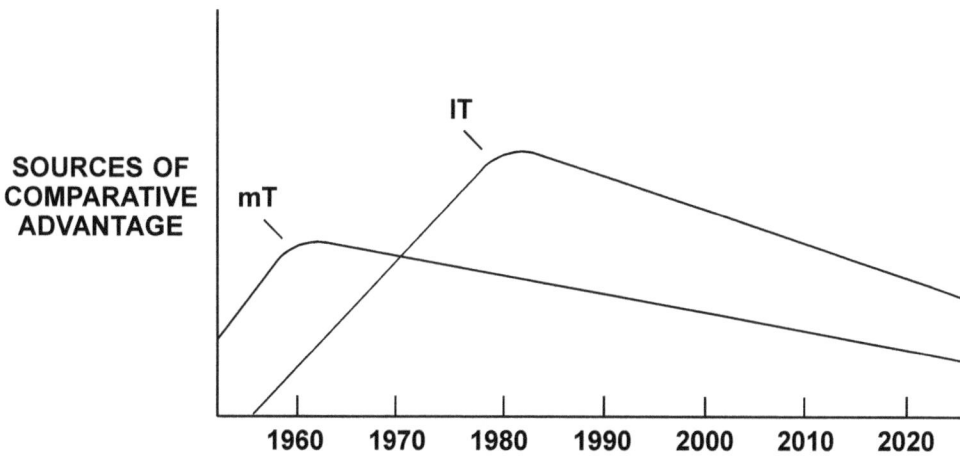

Figure 28. Data-Based Curves of Phenomenal Information

In a similar manner, our projected curves, themselves based upon a multitude of predicted data points, also illustrate phenomenal modeling. As shown in Figure 29, the following technologies rise to become sources of comparative advantage in the market-place: Human Technologies (HT), Organizational Technologies (OT), and Marketplace Technologies (MT). Later, we will discover how these curves represent the introduction of the Age of Ideation.

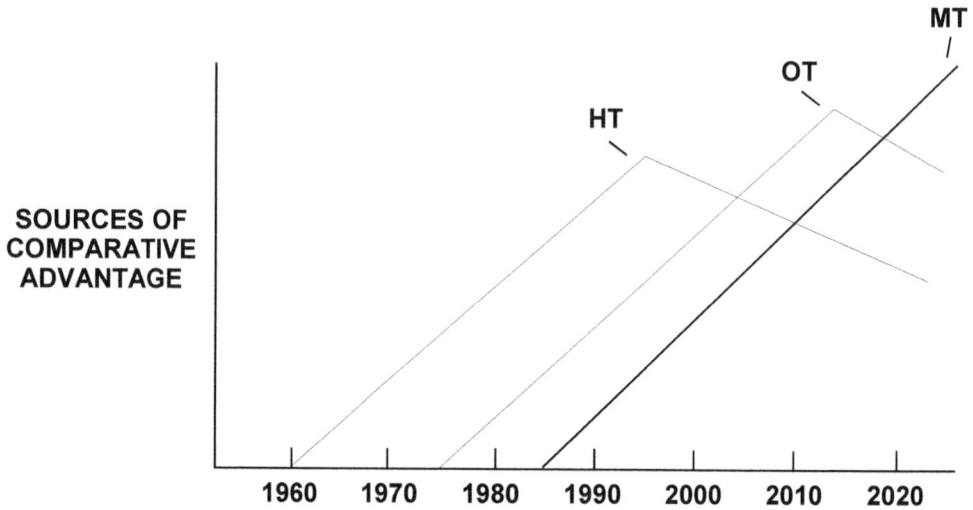

Figure 29. Projected Curves of Phenomenal Information

We may derive vectorial models deductively from phenomenal information as illustrated in Figure 30. Here, multidimensional models are derived and defined operationally to represent the specific vectors that comprise the phenomenal curves. As we can see, lower-order technologies are represented by models that are *nested* in higher-order technologies. Such nesting is an example of representing vectors of phenomena acting upon other phenomena. In this instance, the vectors of one technology act upon other technologies.

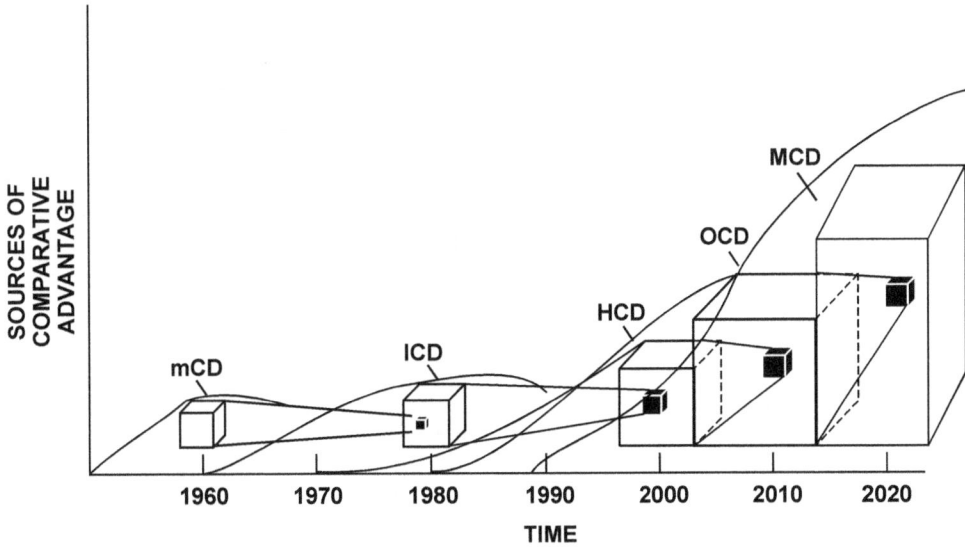

**Figure 30. Vectorial Information Derived from
Phenomenal Information**

The human significance of phenomenal information is that we can derive vectorial models to represent the directions and forces of phenomena; we can model our "universes" as we understand them to be. Within them, we can attempt to derive the multidimensional, operational, and conceptual information that empowers us to develop hypotheses as the basis for interventions of human intentionality.

For measurement and training purposes, levels of representing are scaled by their operations (see Figure 31). As may be noted, the levels are mnemonically presented as the **5Ss:**

LEVELS

1	**Sentences** or conceptual
2	**Systems** or operational
3	**Schematic** or dimensional
4	**Social-Schematic** or vectorial
5	**Spatial** or phenomenal

Each level is nested developmentally in higher-level representations.

LEVELS

OPERATIONS

5. SPATIAL
(Phenomenal)

4. SOCIAL
(Vectorial)

3. SCHEMATIC
(Dimensional)

2. SYSTEMS
(Operational)

1. SENTENCES
(Conceptual)

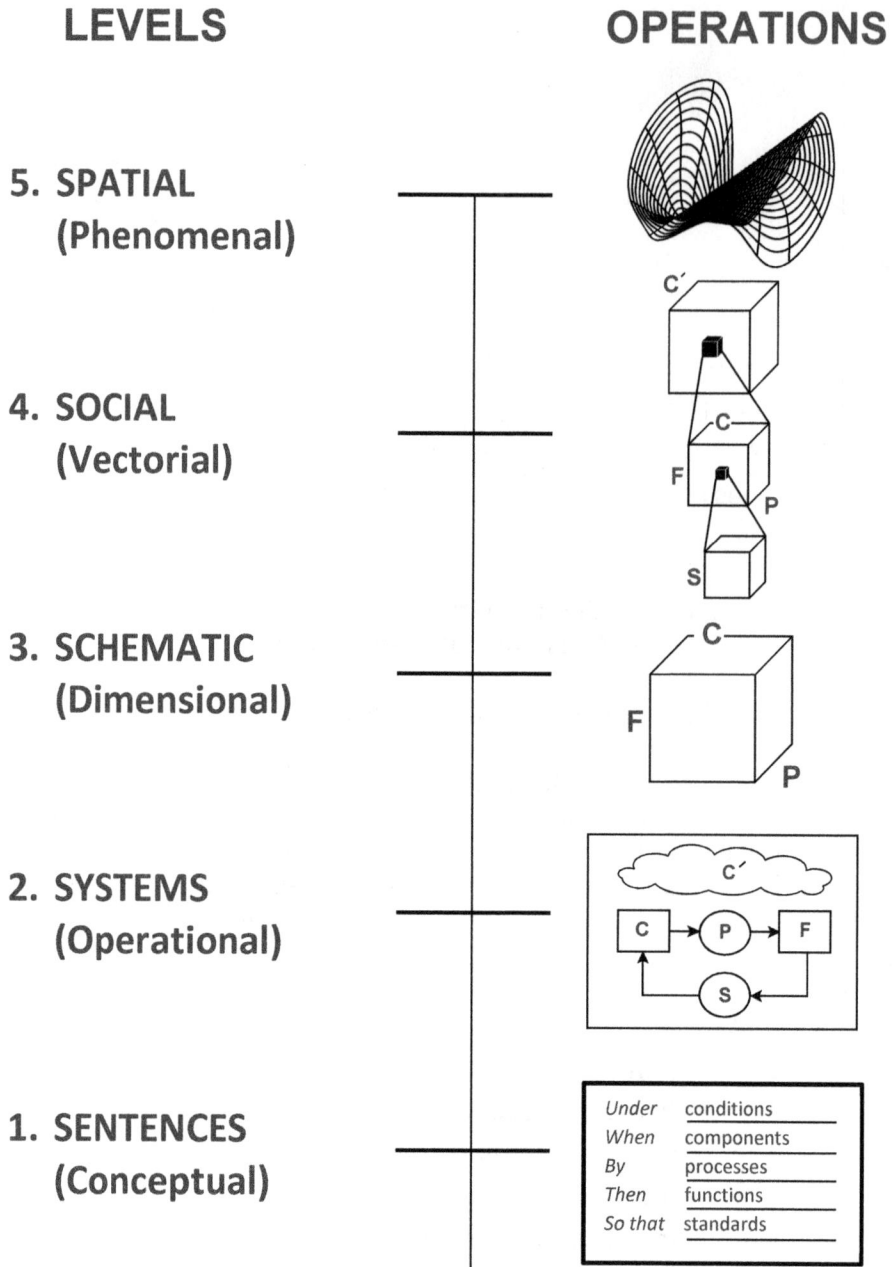

Figure 31. Levels of Representing

5
Human ↔ Information Processing

Generativity is thinking beyond the high-beams. It is as if we were driving along as we use our brainpower in most of our lives—using our low-beams. Once in a while at night, we switch to high-beams and see things that we would never have otherwise seen. This enables us to avoid accidents with the obstacles lying ahead of us.

When we flick on our "thinking beams," we also see beyond the high-beams. This empowers us to see not only things that might have been. This empowers us to see ourselves in relation to things that might be—in other words, possibilities. This empowers us to view the possibilities in our lives. Even in the face of failure, there are exponential degrees of possibilities for success.

It is empowered by interdependent human and information processing (HT ↔ IT). It culminates with a concrete image of information output of actionable plans for operations:

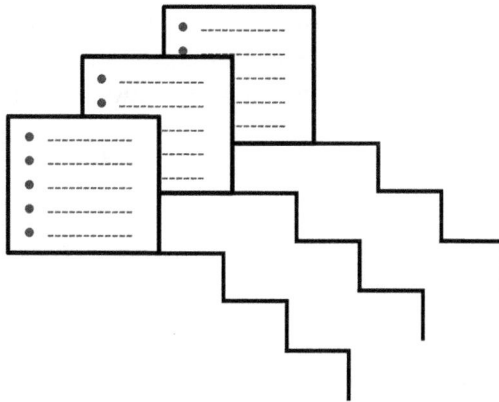

Generativity human and information processing goes through several processing operations to transform the information input into action output. The critical operations are expanding and narrowing. First, we *expand* the possibilities of operations. Then, we *narrow* to the preferred probabilities operation that will define the objectives of our action plan.

This is the simplicity of *generative processing:*

- First, we **relate** to images of information.

- Second, we **represent** these images operationally.

- Third, we **expand** more possibilistic images.

- Fourth, we **narrow** to a preferred probabilistic image.

- Fifth, we **define** an action plan.

Generative processing has been the theme of my life. It is the source from which all breakthrough sciences are generated. It is the force from which all cutting edge technologies are innovated.

Generative processing is the most difficult human experience to articulate: that the raw materials of human experience produce the finished products of human endeavors; that seemingly random information inputs become new capital outputs—individual, organizational, community, cultural, economic; that seemingly "nothing" is made into "something." These are the functions of an information science: to transform raw data inputs into invaluable information capital outputs.

More immediately, our mission is to transform IT users into co-processors:

> **To prepare co-processors for IT programs by empowering them with human generativity systems that process the levels of information representation of the phenomena being addressed**

This is the mission of generativity science for the IT world:

> **Generativity Science = (HC $\leftrightarrow$ IC)**

Generativity science is a function of the synergistic processing relationship between Human and Information Capital: (HC $\leftrightarrow$ IC). Each grows as the other grows.

Goaling by Relating

We initiate all of our generating processing experiences with goaling. In order to define our goals, we must relate to the conditions of our environment. So we must ask and answer the following question:

What are the dimensions of our processing environment?

Extensive factor analysis yields two pre-potent factors upon which all other dimensions load:

1. **Information Representation (IR)**

2. **Human Processing (HP)**

Both factors need to be scaled in order to determine the effects of each level upon every level of both factors (see Table 5).

Table 5. The Pre-Potent Factors of Generative Processing

Information Representing (IR)	Human Processing (HP)
5	5
4	4
3	3
2	2
1	1

In Table 6, we may view the generativity factors in scaled forms. As may be seen, the highest levels of processing nest and encode all of the other levels.

Table 6. The Pre-Potent Factors of Individual Generativity

INFORMATION REPRESENTING (IR)		HUMAN PROCESSING (HP)	
5	Phenomenal	5	Acting
4	Vectorial	4	Understanding
3	Dimensional	3	Exploring
2	Operational	2	Inputting
1	Conceptual	1	Goaling

We may view the levels of operations or IR in further detail in Figure 32. As may be noted, all levels of information operation are "nested" in "Phenomenal" or "Spatial Schematics".

LEVELS	OPERATIONS

5   **PHENOMENALIZING**   **S⁵ – SPATIAL SCHEMATICS**

S^5 – **SPATIAL SCHEMATICS**

5   Standards (S)
4   Conditions (C^1)
3   Processes (P)
2   Components (C)
1   Functions (F)

4   **VECTORIALIZING**

Conditions

Phenomenon

Standards

S^4 – **SOCIAL SCHEMATICS**

5   Standards (S)
4   Conditions (C^1)
3   Processes (P)
2   Components (C)
1   Functions (F)

3   **DIMENSIONALIZING**

S^3 – **SCHEMATICS**

3   Processes (P)
2   Components (C)
1   Functions (F)

2   **OPERATIONALIZING**

S^2 – **SYSTEMS**

5   Standards (S)
4   Conditions (C^1)
3   Processes (P)
2   Components (C)
1   Functions (F)

1   **CONCEPTUALIZING**

S^1 – **SENTENCES**

5   Programs
4   Objectives
3   Principles
2   Concepts
1   Facts

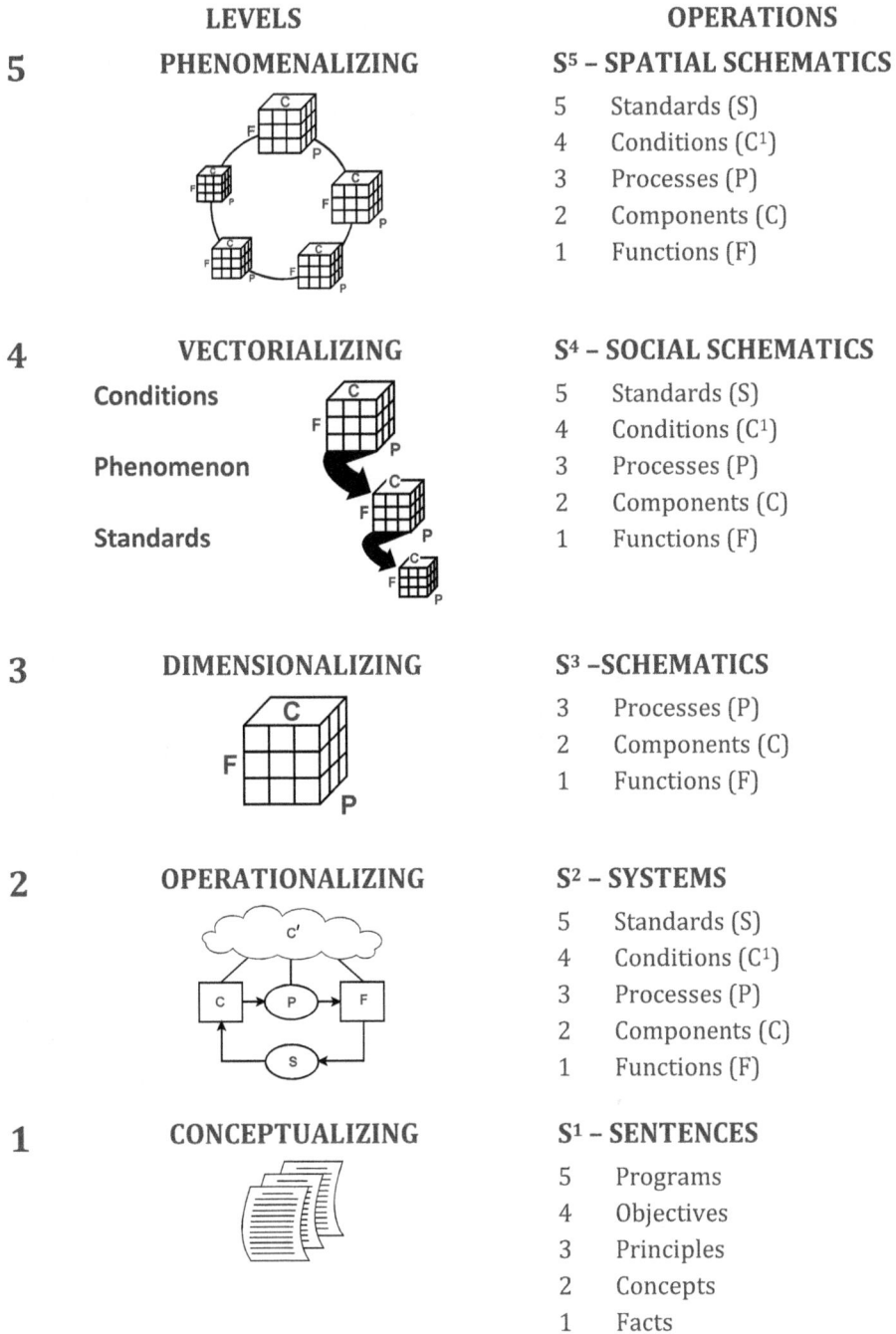

Figure 32. Levels of Information Representing (IR) Operations

We may also review the levels of operations of HP in further detail in Figure 33. As may be noted, all levels of processing operations are nested in "R^5 – Reasoning."

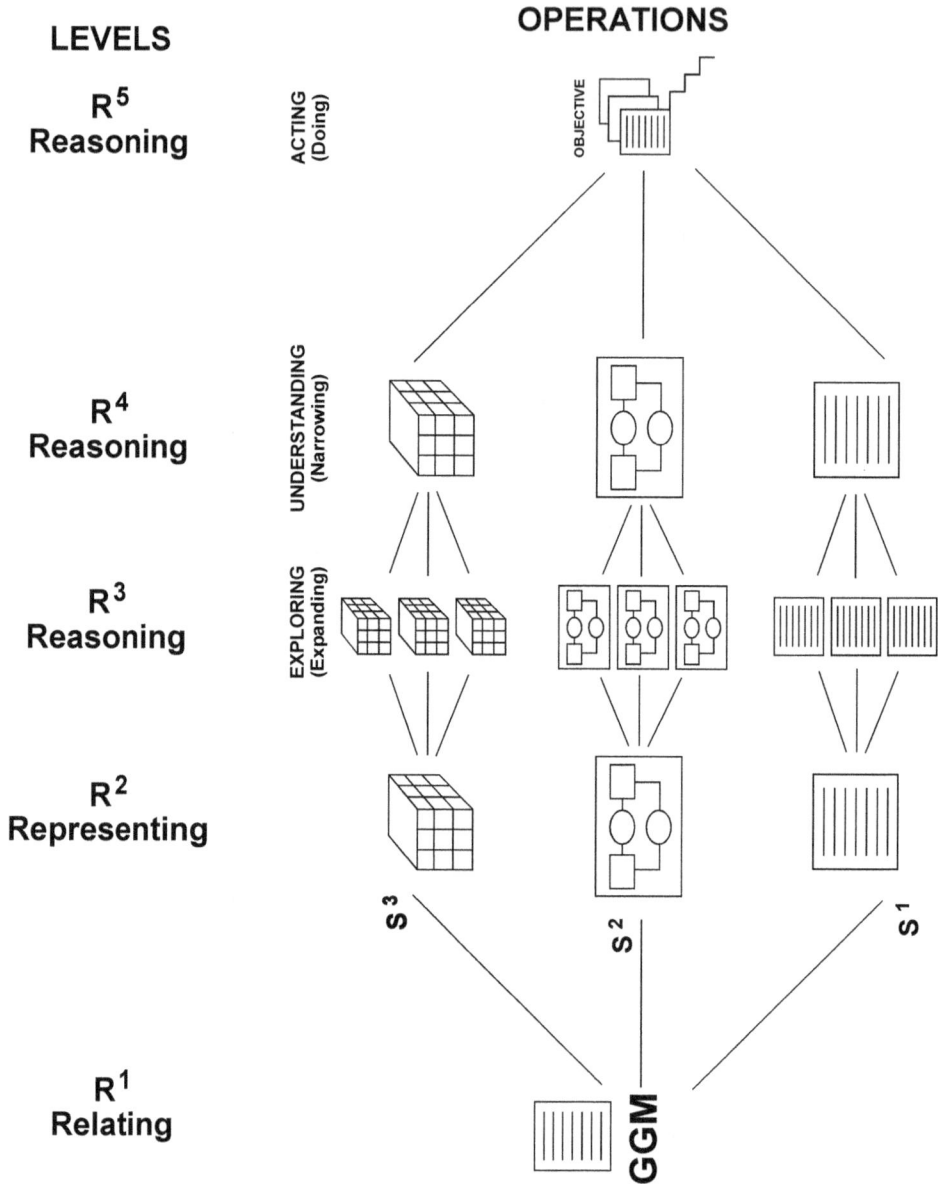

Figure 33. Levels of Human Processing (HP) Operations

Inputting by Representing

Now that we have the factors of individual generative processing, we need to relate them. We relate them inductively, as illustrated in Figure 34, where low levels of HP (1 2 3 4 5) are aligned with low levels of IR (1 2 3 4 5). We also may learn to relate them deductively by aligning high levels of HP (5 4 3 2 1) with high levels of IR (5 4 3 2 1). This matrix defines the boundaries of the Human-Information Universe in which we are processing. This Human-Information Generativity Matrix will tell us the effects of each level upon every other level of both factors (see Figure 34).

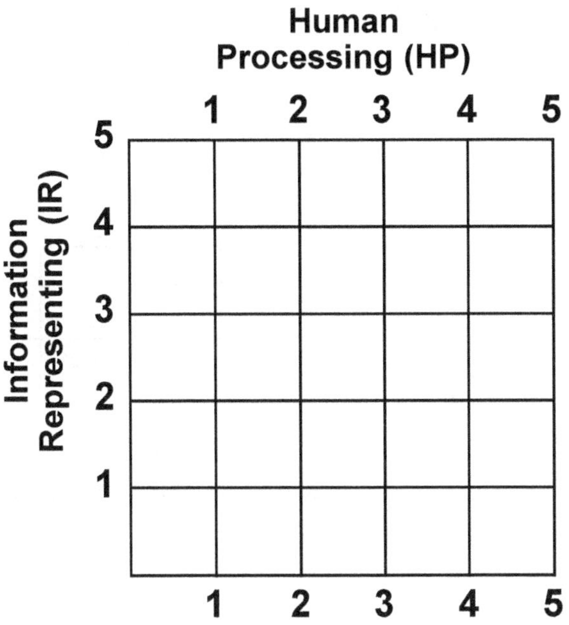

Figure 34. The Human-Information Generativity Matrix (Inductive)

In Figure 35, we may view the generativity matrix deductively with high levels of HP with high levels of IR. As may be seen, all phases of HP process all levels of IR.

IR \ HP	5. Goaling	4. Inputting	3. Exploring	2. Understanding	1. Acting
5. Phenomenal					
4. Vectorial					
3. Dimensional					
2. Operational					
1. Conceptual					

Figure 35. The Human-Information Generativity Matrix (Deductive)

Exploring by Expanding

Now that we have defined the universe of human achievement, we can initiate human processing or reasoning. The first stage in Reasoning, or R^1, is exploring by expanding (see Figure 36). As may be noted, expanding emphasizes analyzing the highest levels of IR. This satisfies our values of perspective on the lower levels of information representation. We go as high on IR as is necessary to understand the phenomena we are addressing. In other words, we go as high as our values dictate.

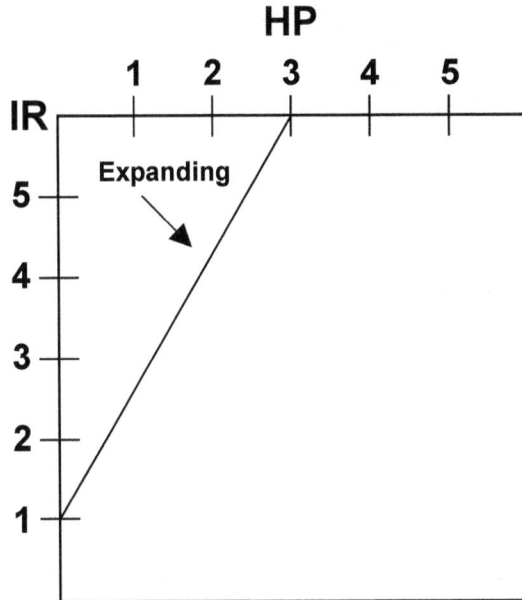

Figure 36. Exploring by Expanding Levels of IR

In Figure 37, we may view the generativity matrix deductively by converging with high levels of HP and IR. As may be seen in the illustration, initially low levels of IR (conceptual) are expanded to meet our values of perspective of IR (vectorial).

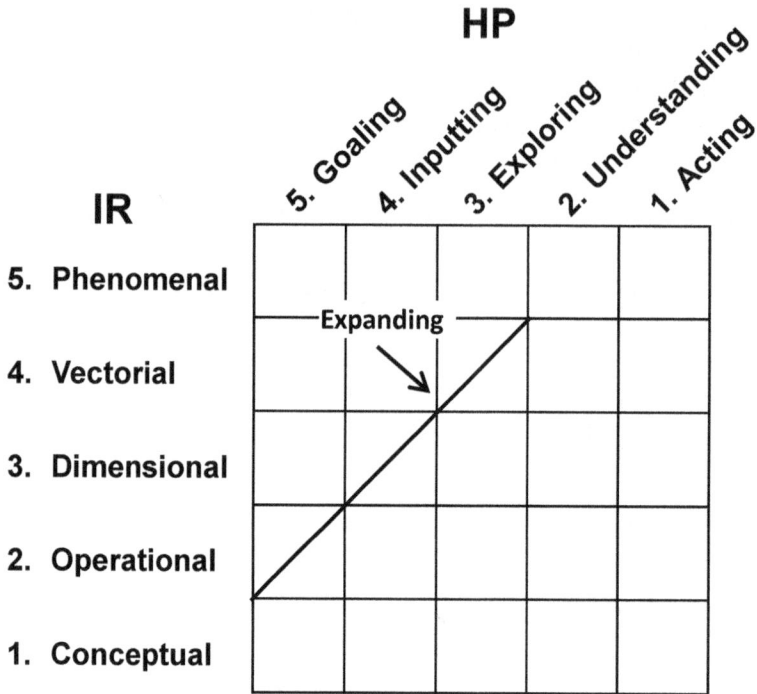

HP

IR

5. Goaling
4. Inputting
3. Exploring
2. Understanding
1. Acting

5. Phenomenal

Expanding

4. Vectorial

3. Dimensional

2. Operational

1. Conceptual

Figure 37. Exploring by Expanding IR Levels (Deductive)

Understanding by Narrowing

Now that we have our greatest perspective, we seek to narrow our objectives by defining the conditional requirements. In other words, just as we view our universe with our values, so does our universe view us with its values: **Do we meet its requirements?** If so, we can go on to define the operations of our objectives (**O**) (see Figure 38).

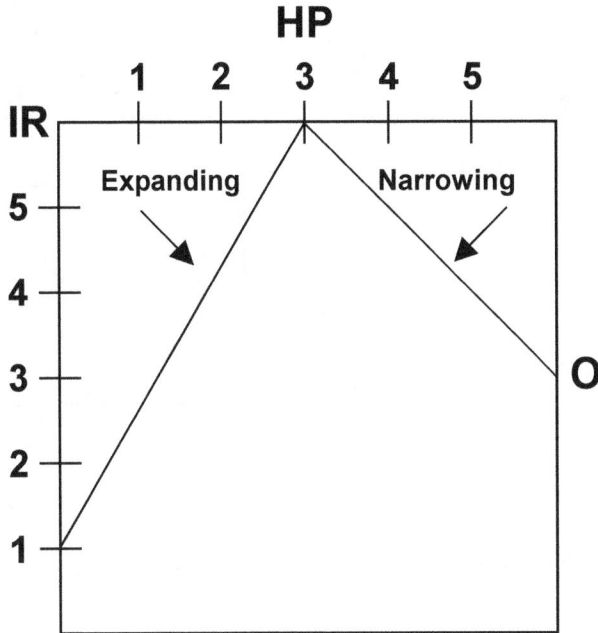

Figure 38. Understanding by Narrowing to IR Requirements

In Figure 39, we may again view the generativity matrix deductively. As may be seen on the illustration, transitionally high levels of IR (vectorial) are narrowed to moderate levels of IR (dimensional) in order to meet requirements.

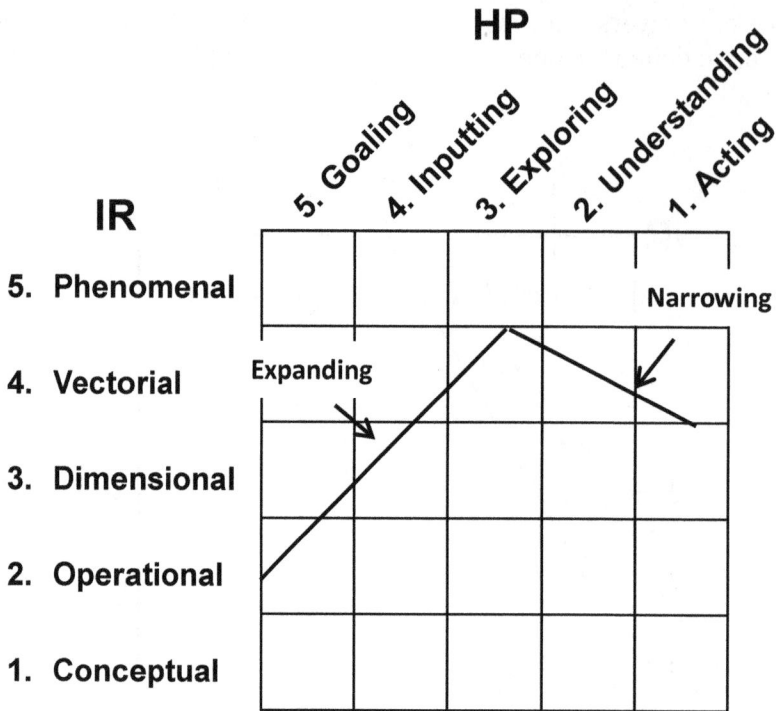

Figure 39. Understanding by Narrowing to IR Requirements (Deductive)

Acting by Objectifying

Finally, having narrowed on our target objectives (O), we can now define them operationally—O^1, O^2, O^n (see Figure 40). In other words, we can define the Os in operational dimensions that our narrowing operations dictate:

- **Functions**—responses or products
- **Components**—parts or participants
- **Processes**—procedures or methods
- **Conditions**—contexts or environment
- **Standards**—achievement or excellence

These operational definitions make our targeted Os actionable and achievable.

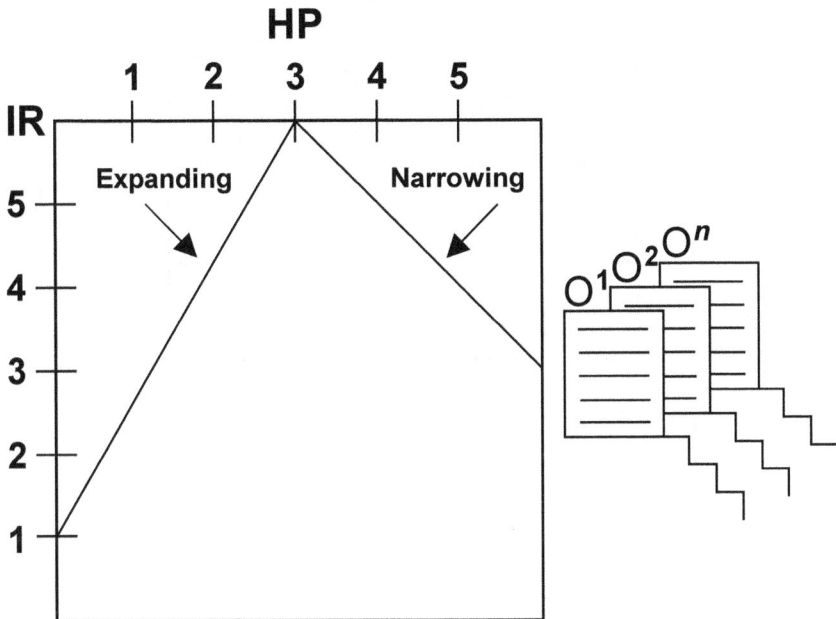

Figure 40. Acting by Operationalizing Targeted Objectives

In Figure 41, we may view the generativity matrix deductively. As may be seen in the illustration, the HP culminates in targeted objectives that are defined in operational and therefore actionable terms.

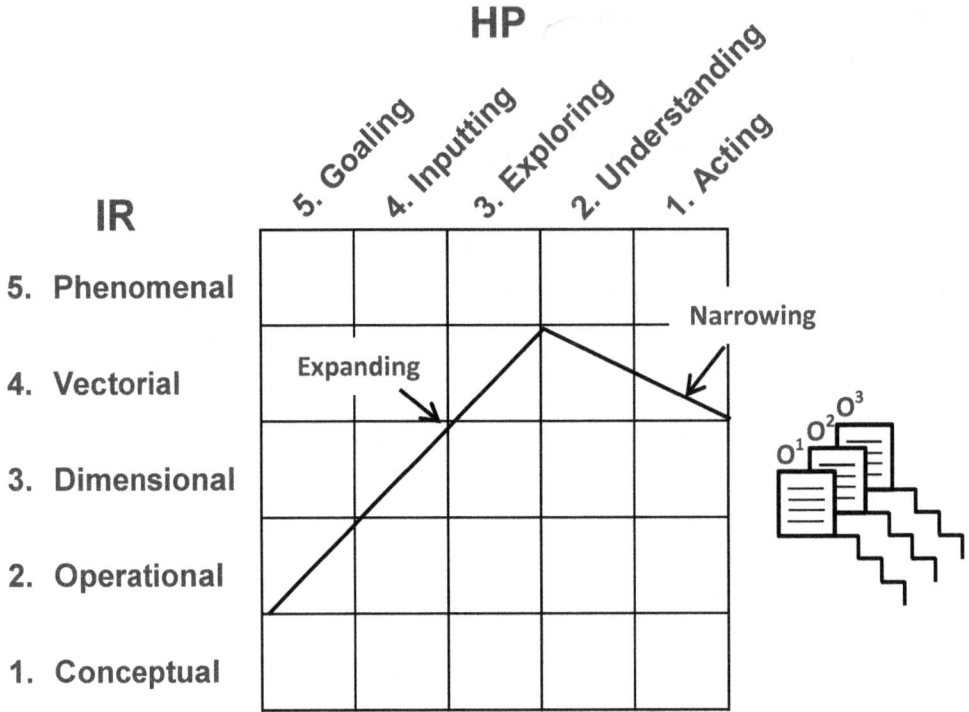

Figure 41. Acting by Operationalizing Targeted Objectives (Deductive)

In transition, we began by relating to determine our goals for processing and developing our generativity matrices. We concluded by operationally defining our targeted objectives (see Figure 42). Transitionally, we expanded, narrowed, and prepared for acting upon these objectives. These are the ingredients of individual generativity:

- **Goaling** by relating
- **Inputting** by representing
- **Exploring** by expanding
- **Understanding** by narrowing
- **Acting** by operationalizing

In the pages that follow, we will apply our cubing technologies. We will discover the potentially infinite potential of information cubing.

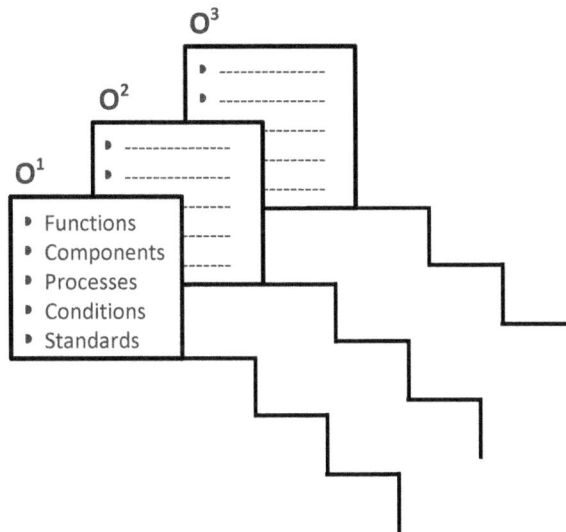

Figure 42. Operationalizing Objectives

IV
Information Generativity

6
Possibilities → Information

The old scientific components emphasized conceptual knowledge: facts, concepts, principles. It was assumed that this conceptual and operational information would enable the discharge of the old scientific functions: description, prediction, control. Wrong! You can't get there from here! Indeed, you never could! That belief is a vestige of an artifactual science. We never could explain humanly meaningful phenomena in terms that would control their variability. There is simply too much information capital remaining to be developed.

The new scientific components emphasize the expanded information components. Not only is the conceptual component completed, but all qualitative components of information modeling are added. In model-building terms, we may say that each information component is *nested* as a cell in a higher-order component. The new science is exponentially more powerful in discharging the freeing function of changeability.

In short, the new science is truly an information science. In reality, it is the first information science because it is the first to develop information capital in systematic ways. The current information technologies—no matter how powerful their hardware or ubiquitous their software—are incapable of meeting information requirements without comprehending the power of information capital development. Connecting—even communicating, collaborating, and coordinating—does not yield *capitalizing*. We must reflect upon how powerfully informed we are by this ICD processing; and we have not yet fully begun to process interdependently!

The Probabilities Problem

The possibilities scientist realizes that in the context of **ICD,** the old science is highly problematic. Even with its consideration of randomness and chance, the utility of linear or even curvilinear classification of variables, and the careful standard definitions of method, probabilities science employs time-fixed samples for data: the independent variables are confined; the intervening variables are controlled within rigid limits; and the dependent variables are *"freed"* only under controlled conditions.

All of these characteristics are measured with *"scales"* based upon predetermined suppositions, none of which are responsive to changing possibilities. Even so-called *"degrees of freedom"* are the products of fixed, controlled, and balanced equations—the degrees are never *"free."* They are never at liberty to reflect empowered, free phenomena that are processing still more possibilities. In short, the probabilities suppositions are based upon the most elusive assumptions of what is normal and what is skewed in two fixed dimensions, *fixed in time!*

When probabilities science ventures into a third dimension, it represents measurement with straight lines connected to two or more other straight lines, all at right angles to one another. There is nothing more restrictive and controlling, and therefore less empowering, than straight lines meeting to form right angles—and, we may add, curved lines reflecting time-fixed samples, controlled intervening variables, rigid intervening variables, and the magnitude of anything but free dependent variables.

Organizational Alignment

The highest level of information modeling reflects the highest level of inter-dependency: phenomenal information components discharging phenomenal freeing functions by interdependent phenomenal processing systems, or I^5. We label this *phenomenal intelligence* because it defines and relates the universes of the phenomena involved—it frees phenomena by processing all phenomena to seek their own unique and changeable destinies. In doing so, phenomenal intelligence initiates deductive processing.

At the highest levels, we can see this intelligence in operation in its transformation into phenomena. At such levels, our source model, possibilities science, yields its first and highest-level source cell—the phenomenal intelligence cell. This cell positions phenomenal information components to accomplish phenomenal freeing functions by interdependent phenomenal processing systems.

In this context, we may view deductive intelligence in operation by modeling marketplace-driven vectorial systems. An overview of areas of **New Capital Development (NCD)** systems to be processed includes

- **MCD**—marketplace capital development;
- **OCD**—organizational capital development;
- **HCD**—human capital development;
- **ICD**—information capital development;
- **mCD**—mechanical capital development.

In the illustrations that follow, the NCD system deductively models the alignment of all systems: MCD, OCD, HCD, ICD, mCD.

MCD

In the marketplace capital development model (Figure 43), the MCD functions, components, and processes are identified. The functions of MCD are derived from the market's requirements for NCD systems: MCD, OCD, HCD, ICD, and mCD. In other words, the marketplace of organizations is dedicated to fulfilling these NCD requirements. The MCD components are the corporate technologies available among organizations in the marketplace: MT, OT, HT, IT, and mT. These technologies are critical to fulfilling the requirements of the marketplace. The processes of the MCD model are organizational processing systems: leadership, marketing, resources, technology, and production.

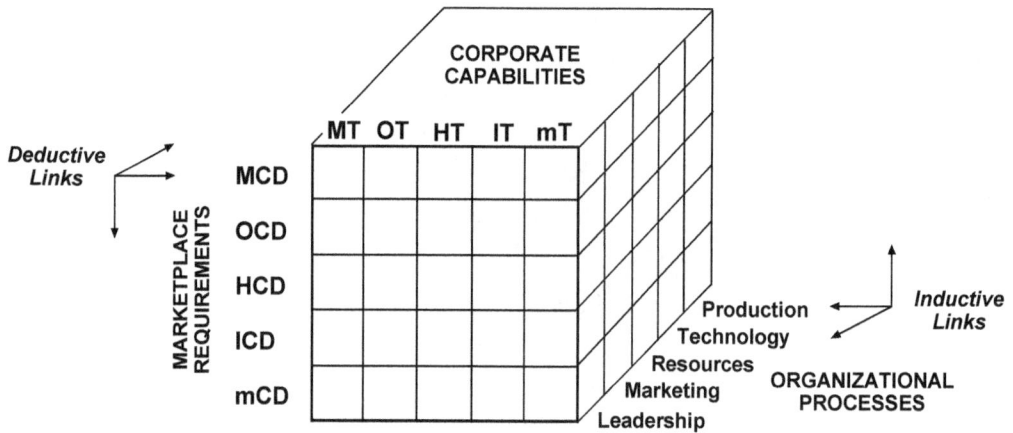

Figure 43. Marketplace Capital Development Model

To summarize:

Marketplace technologies are dedicated to new capital development enabled by organizational processing systems.

OCD

The functions of OCD are derived from the market's technology requirements, as shown in the model below (Figure 44). These requirements are translated operationally into functional levels of the organization: policy, executive, management, supervision, and delivery. In other words, the resources of the organization will be dedicated to fulfilling these marketplace requirements. The OCD components are units of the organization and are derived from the processes of the MCD model: leadership, marketing, resources, technology, and production. These organizational units are critical to fulfilling market requirements. The processes of the OCD model are introduced as HCD processes: goaling, inputting, processing, planning, and outputting. These HCD processes are essential for organizations to fulfill their goals.

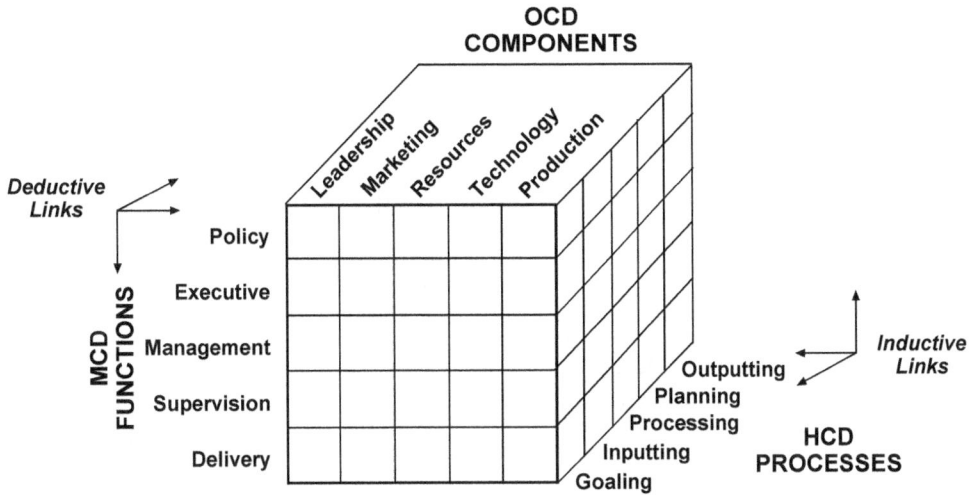

Figure 44. Organizational Capital Development Model

To summarize:

OCD components are dedicated to MCD functions enabled by HCD processing systems.

HCD

As may be noted in the human capital development model (Figure 45), OCD components have been rotated to become the functions of HCD: leadership, marketing, resources, technology, and production. In other words, human capital is dedicated to fulfilling these organizational goals. Similarly, the HCD processes of the OCD model have been rotated to become HCD components: goaling, inputting, processing, planning, and outputting. These human processing components are critical to fulfilling the goals of the organization. Finally, the ICD processes by which the HCD components discharge OCD functions are introduced: phenomenal, vectorial, dimensional, operational, and conceptual. These ICD processes are essential for human processing.

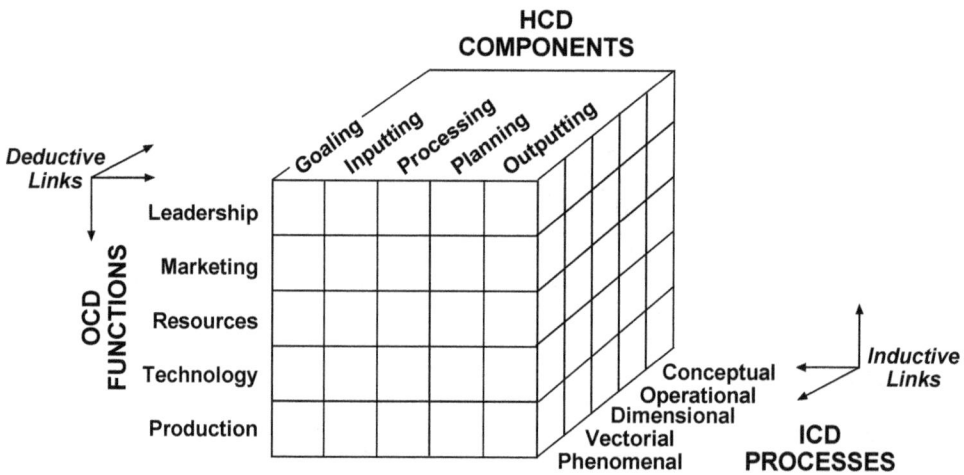

Figure 45. Human Capital Development Model

Again, note that each lower-order ingredient is dedicated to enabling the achievement of a higher-order function:

HCD components are dedicated to OCD functions and enabled by ICD processes.

ICD

In the information capital development model (Figure 46), HCD components have been rotated to become the functions of ICD. In other words, information capital is dedicated to servicing the requirements of thinking people. Likewise, the ICD processes of the HCD model have been rotated to become ICD components. These information components are critical ingredients in the service of human processing. Finally, mCD operations are introduced as enabling operationalizing processes: functions, components, processes, conditions, and standards. These mechanical operations are essential to information capital processing.

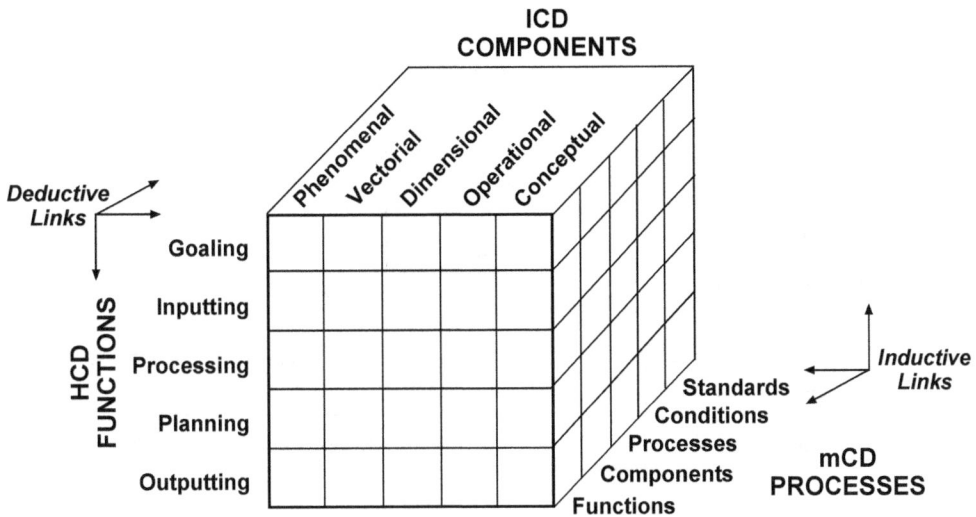

Figure 46. Information Capital Development Model

Again, note that lower-order ingredients are dedicated to achieving higher-order functions:

ICD components service HCD goals through mCD processes.

71

mCD

The mechanical capital development model (Figure 47) has ICD components that have been rotated to become the functions of mCD. Mechanical capital or mechanical tools are dedicated to service information designs or ICD. In turn, mCD processes of the ICD model have been rotated to become mCD components. These mechanical components are critical to fulfilling information designs. Finally, new mCD programming processes (mCD') are introduced: programs, instructions, tasks, steps, and implementation. These programmatic mechanical processes are essential to mechanical processing.

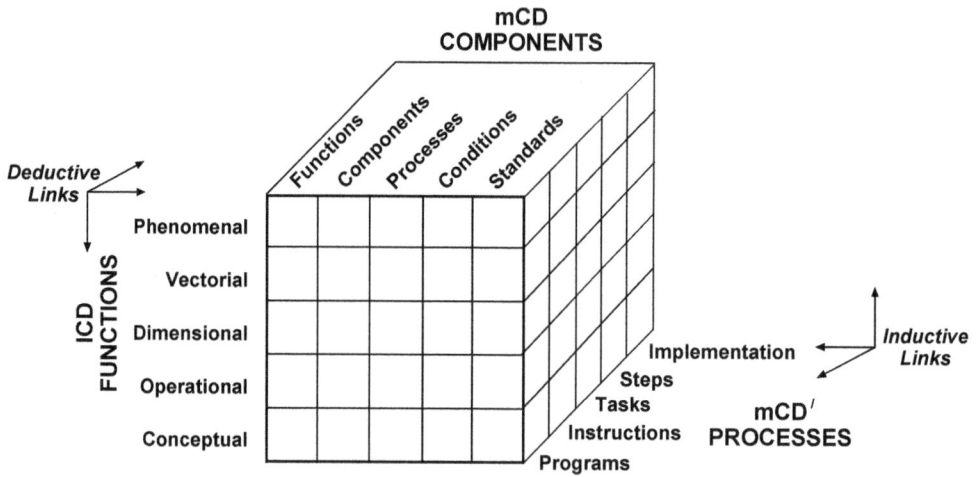

Figure 47. Mechanical Capital Development Model

Again, lower-order ingredients are dedicated to higher-order functions:

mCD components service ICD goals through mCD' processes.

Sourcing from the MCD model, then, we deduce our remaining models vectorially: OCD, HCD, ICD, mCD (see Figure 48). Each lower-order processing system is *nested* in a higher-order processing system. All are *nested* in the MCD model. Moreover, each lower-order model is deduced, or derived, by rotating components and processes from the higher-order source. All are transformed in identity as we derive the level of their functions. This is deductive intelligence in operation.

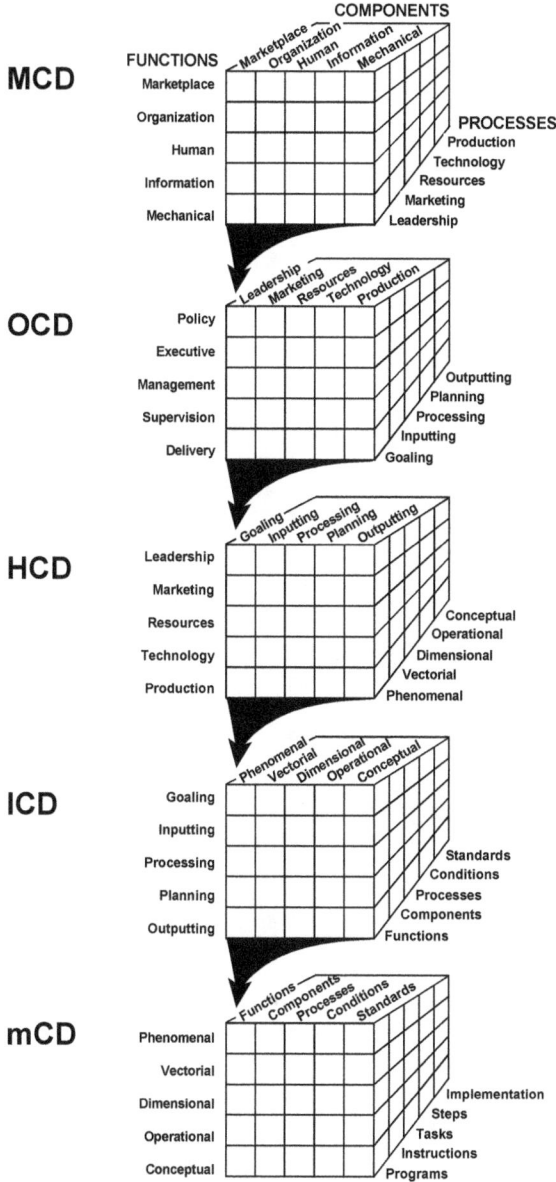

Figure 48. Deductive Intelligence

7
Information "Cubing"

The net of our processing of phenomenal experience may be viewed in our *information "cubing"* phenomena: phenomenal functions, components, and processes (see Figure 49). As may be noted:

- Phenomenal functions are driven by the discovery of new phenomena.

- Information components are driven by phenomenal representations.

- Human-information processes are driven by the phenomenal processing systems (S–PP–R).

We may summarize the result of Information Cubing operationally:

Phenomenal functions are achieved by information components empowered by human-information processing.

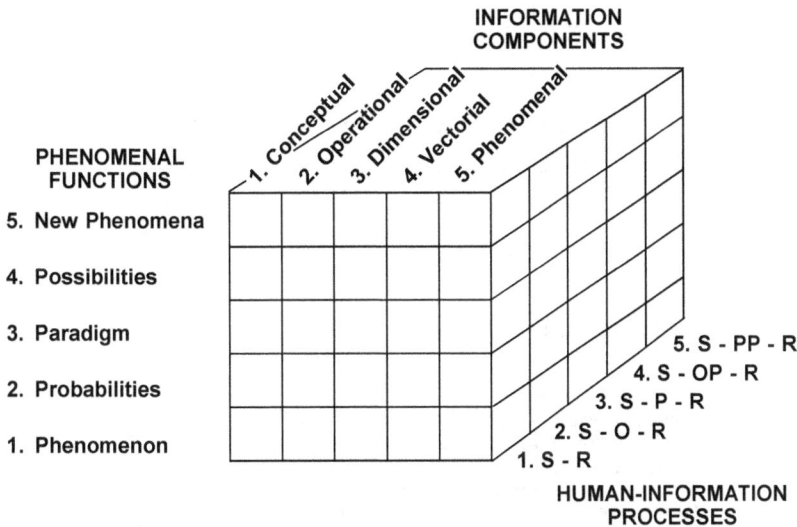

Figure 49. "Cubing" Phenomena

We are now empowered to generate the phenomenal conditions within which the phenomenon is "nested" or "housed" (see Figure 50):

- **Generativity functions** are driven by freeing or releasing functions.
- **Phenomenal components** are driven by new phenomena.
- **Information processes** are driven by phenomenal images.

Again, we may summarize the operational definition of conditions resulting from **"Information Cubing":**

> **Generativity functions are achieved by phenomenal components empowered by information processes.**

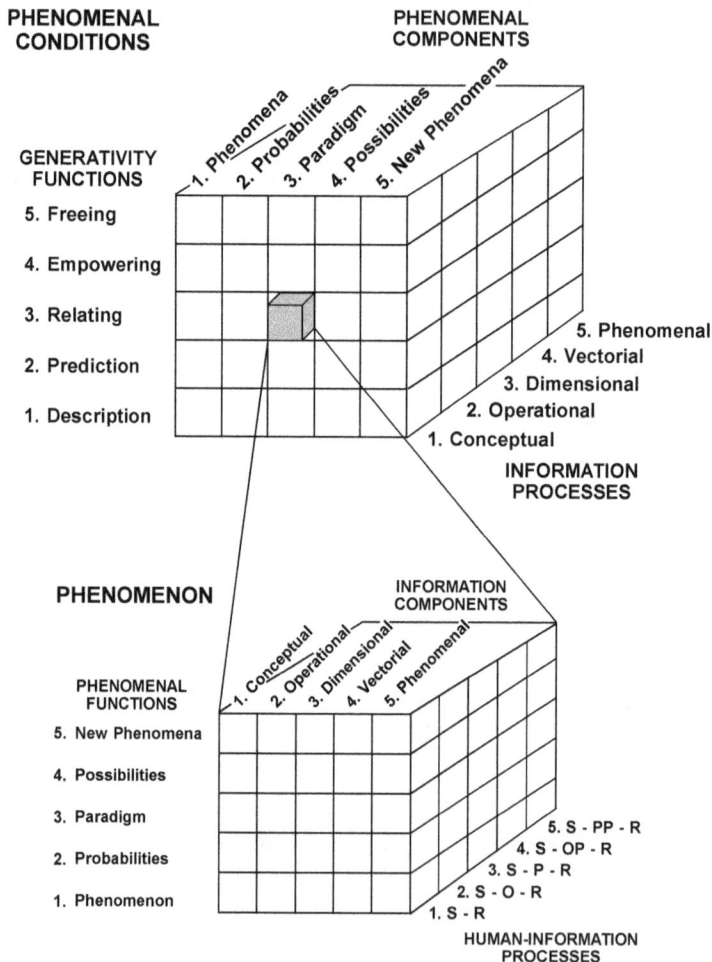

Figure 50. "Cubing" Phenomenal Conditions

Finally, we can generate the phenomenal standards by which we measure our level of achievement (see Figure 51):

- **Information functions** are driven by phenomenal representations.

- **Human-information processing components** are driven by phenomenal processing systems (S–PP–R).

- **Mechanical processes** are driven by phenomenal performance standards.

Once again, we may summarize the operational definition of standards resulting from cubing:

Information functions are achieved by human-information processing components empowered by mechanical processes.

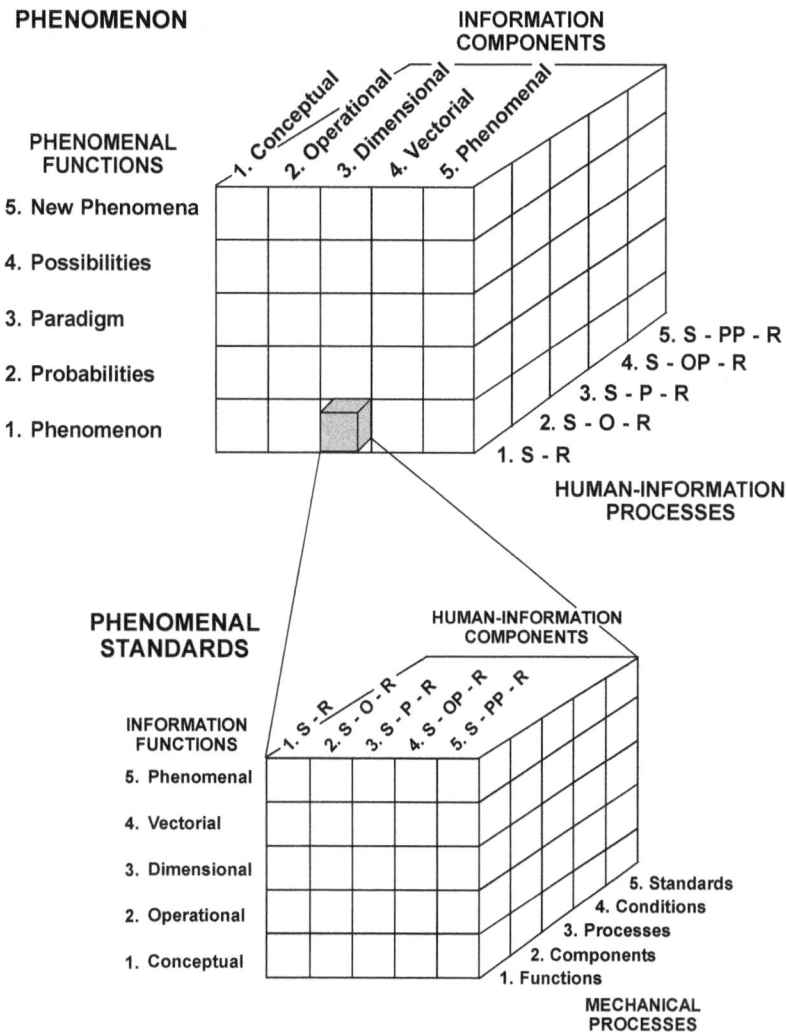

Figure 51. "Cubing" Phenomenal Standards

Information Is Living

The fundamental difference between living and non-living things is information. Living things use information to birth, maintain, create, reproduce, and even terminate themselves. Nonliving things do not engage in any of these processes.

Information becomes potent when it is encoded in sequences of symbols such as alphabetical letters, musical notes, or some physical forms that are decoded by books and symphonies or some form of machinery. In short, information needs machinery that, in turn, needs information.

Here is how life works (see Figure 52). One chain of molecules is the DNA that carries information. The other chain of amino acids is linked to proteins and carries on the business of life—birthing, growing, sustaining, reproducing. Notably DNA is not a "blueprint" but rather like a recipe with instructions to be followed.

In human cells, the DNA resides in 46 double-stranded chromosome chains. The chains are super-long and super-thin, reaching millions of miles, yet fitting into the nuclei of each of our trillions of cells.

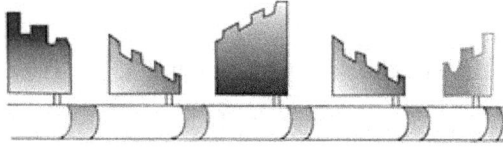

Information chains (DNA and RNA) made of four units (nucleotides)

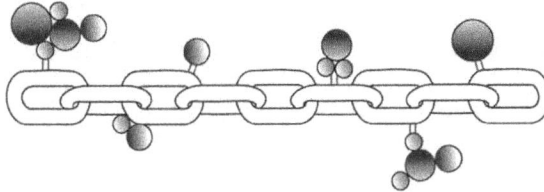

Working or structural chains (proteins) made of 20 units (amino acids)

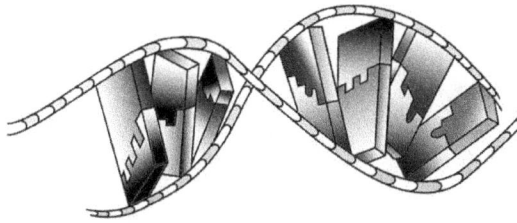

For protection, easy access, and duplication,
DNA chains twist into a double helix.

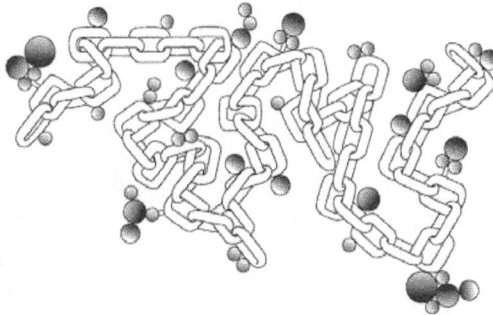

For proper functioning, proteins fold into complicated shapes.
In this way, two-dimensional chains become three-dimensional machinery.

Figure 52. How Life Works (Hoagland and Dodson, 1995)

The process by which the chains of molecules are empowered in a "nesting" of nucleotides (see Figure 53):

- **Nucleotides** are the smallest information units.

- **Genes** comprise nucleotides that specify proteins.

- **Chromosomes** are spooled strings of genes housed in a single unit.

- **Genomes** are collected in the nucleus of each of their cells.

The four nucleotides of DNA comprise the letters of its language of heredity:

- **A** – adenylic acid
- **T** – thymidylic acid
- **C** – cytidylic acid
- **G** – gaunylic acid

Each is a unique arrangement of carbon, nitrogen, oxygen, and hydrogenations called a base.

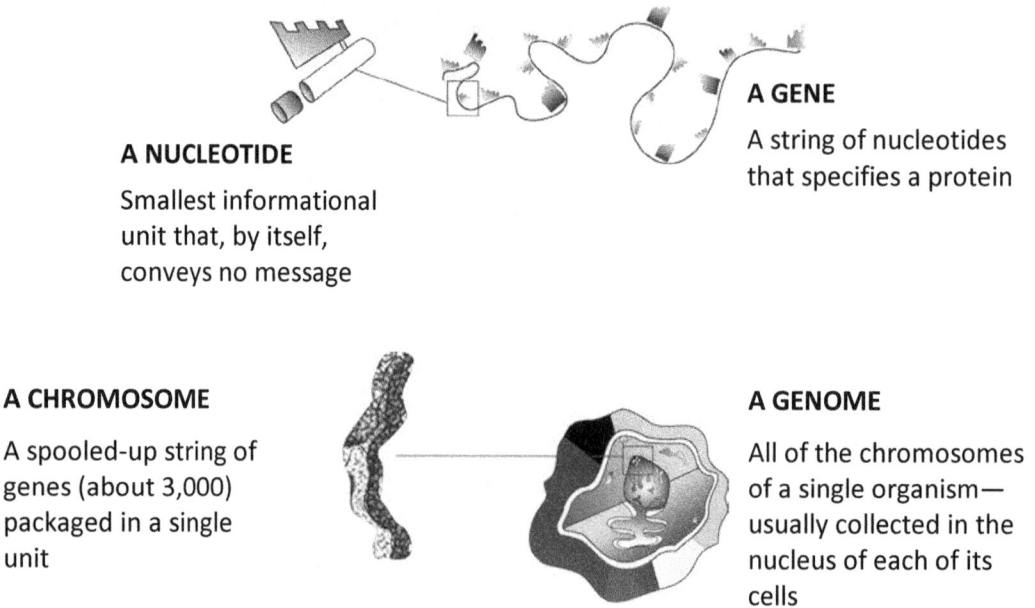

A NUCLEOTIDE

Smallest informational unit that, by itself, conveys no message

A GENE

A string of nucleotides that specifies a protein

A CHROMOSOME

A spooled-up string of genes (about 3,000) packaged in a single unit

A GENOME

All of the chromosomes of a single organism— usually collected in the nucleus of each of its cells

Figure 53. The "Nesting" of Nucleotides

"Cubing" Life

Just as we cube all other phenomena in human experience and human endeavors, we may cube DNA information (see Figure 54). As may be viewed, the following phenomena are represented:

- **Life functions**
- **DNA components**
- **Nucleotidal processes**

We may define its operational definition as follows:

Life functions are achieved by DNA components empowered by Nucleotidal processes.

Cubing these life operations empowers human generativity. The human is not simply the dependent recipient; the human becomes a collaborative partner with the very life experiences that are impacting him or her. The human becomes the interdependent and synergistic generator with life itself: each grows as the other grows.

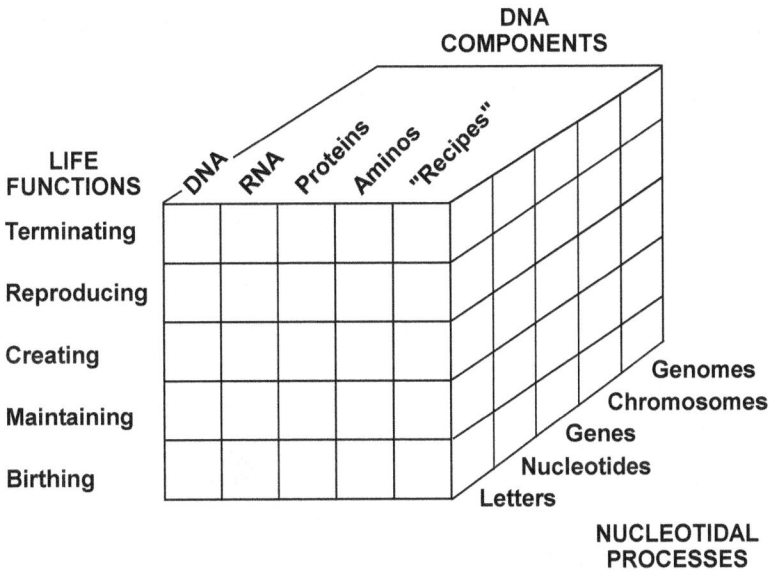

Figure 54. "Cubing" DNA Phenomenal Information

We can generate the very source of DNA by rotating the DNA phenomenon inductively or clockwise (see Figure 55). As may be noted, the Life Components are now dedicated to the Generativity Functions and empowered by the DNA Processes.

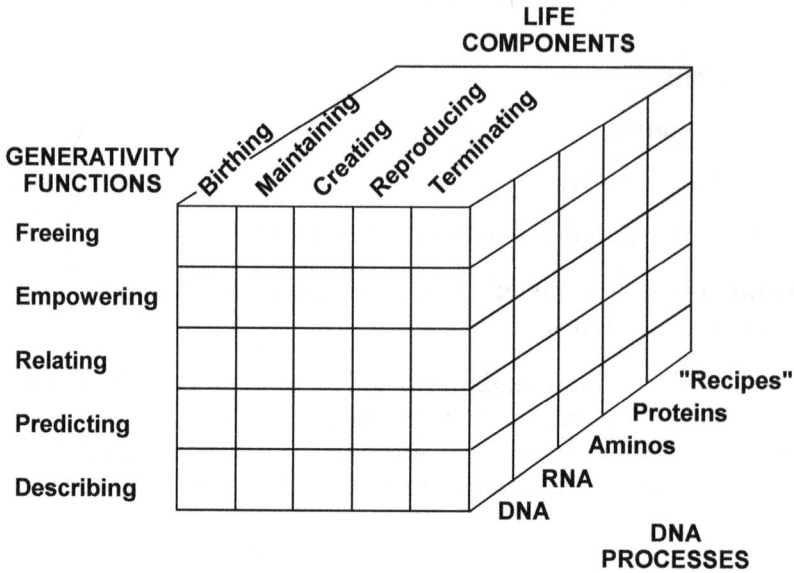

Figure 55. "Cubing" DNA Phenomenal Conditions

Similarly, we can generate the performance standards of DNA by rotating DNA phenomena deductively or counter-clockwise (see Figure 56). As may be noted, the nucleotide components are now dedicated to DNA Functions and empowered by mechanical processes: functions, components, processes, conditions, standards.

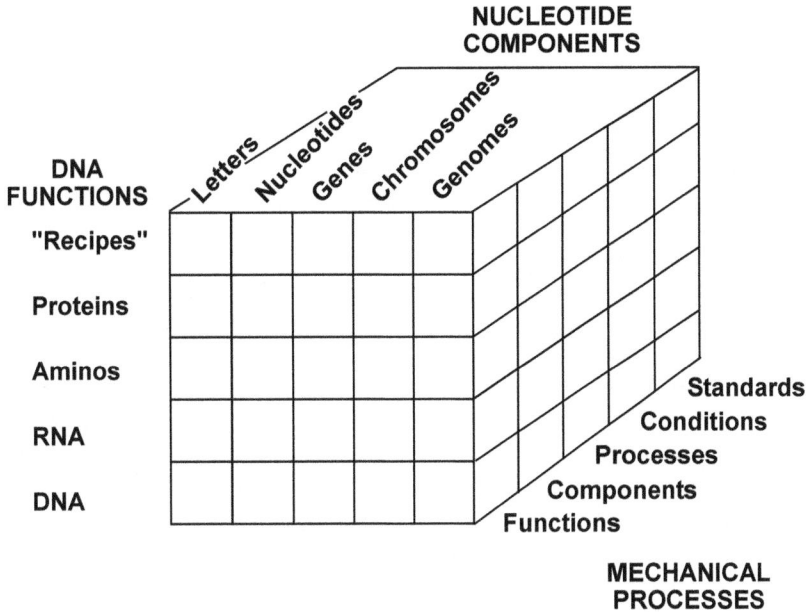

Figure 56. "Cubing" Performance Standards

As may be noted, the human processors may induce the very generativity that will guide them through life. By inductively rotating (clockwise) to the Generativity Functions to which the Life Components were dedicated, the human processors are empowered to generate their life's missions:

- To relate interdependently

- To empower intentionally

- To free intellectually

Now we can ask the critical questions of life:

- How can we become synergistic with life's experiences?

- What are the resource and skill level requirements?

- What are the degrees of freedom and enterprise required?

Modeled upon DNA, information cubing empowers us to model all of life's operations, measure our performance, and project our future opportunities.

Information cubing is the source of life itself!

V

Generating "Something More..."

8
"Nothing But..."

While all expressions of all operations—natural or otherwise—constitute information in the conventional sense of the message being conveyed, there is the realm of information potential that we can generate. In a very real sense, we are blessed with the capacity to live in the world we create.

At the most basic level, there are two orientations to science:

- **"Nothing but..."** or the reductionistic operations with which engineers work

- **"Something more..."** or the expansionistic concepts with which scientists— applied as well as theoretical—work

Succinctly, the operational methods deliver measurable outcomes while the ideational concepts generate testable hypotheses.

Once again, these are two different orientations to generating scientific break-throughs:

- **Inductively,** by moving from operational methods to ideational concepts

- **Deductively,** by moving from ideational models to testing operational hypotheses

The truth of generative science is this: it is only when we integrate the inductive and deductive processing that science works in tandem. We label this **"Hypothetical Model-Building."**

When we initiated our architecture of human-centric software, we began with our **"Nothing but..."** operational definition of our source model:

> **Interdependent functions are achieved by multidimensional components enabled by human processing processes.**

Such an operational definition empowers us to define information models multidimensionally (see Figure 57).

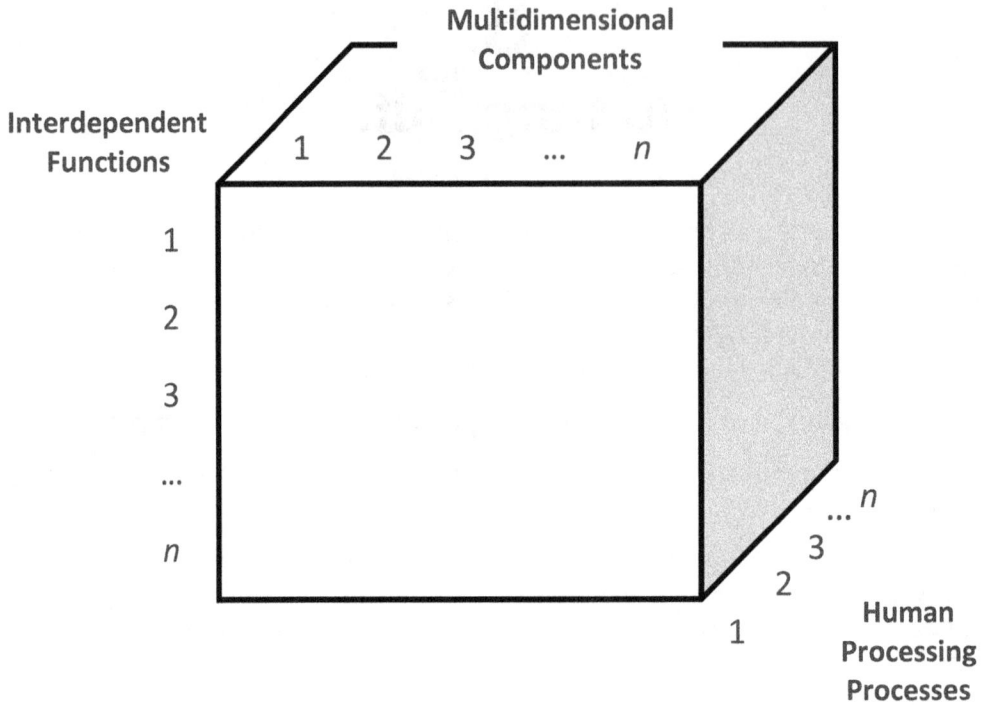

Figure 57. Operational Information Model for Phenomena

We know that our operational definition is incomplete without including the conditions within which the phenomena operate:

Under generativity conditions, generativity functions are achieved by interdependent components enabled by multi-dimensional processes.

We have now defined the conditions within which the phenomena are "nested" by rotating the phenomenal dimensions inductively or clockwise and dedicating them to generativity functions (see Figure 58).

CONDITIONS

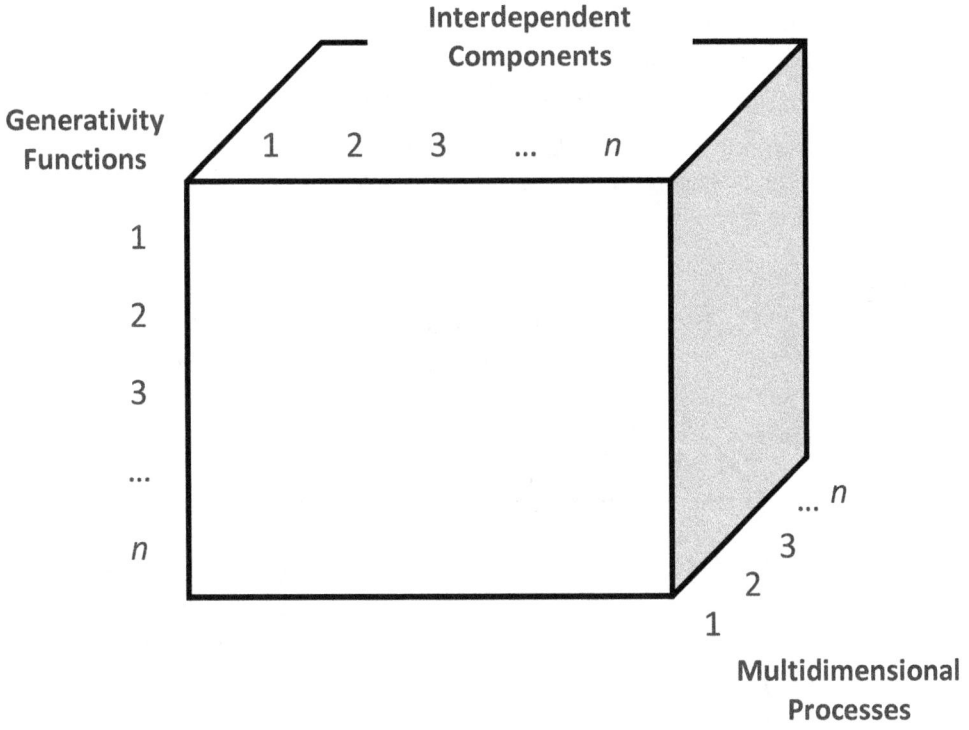

Interdependent Components

Generativity Functions

1   2   3   … n

1

2

3

…

n

… n
3
2
1

Multidimensional Processes

Figure 58. Operational Information Model for Conditions

We also know that our operational definition does not culminate without the standards of performance:

Under generativity conditions, interdependent functions are achieved by multidimensional components enabled by human processing processes at measurably changeable levels.

We have now defined the standards by rotating the phenomenal dimensions deductively or counter-clockwise and dedicating them to multidimensional changeability functions (see Figure 59).

STANDARDS

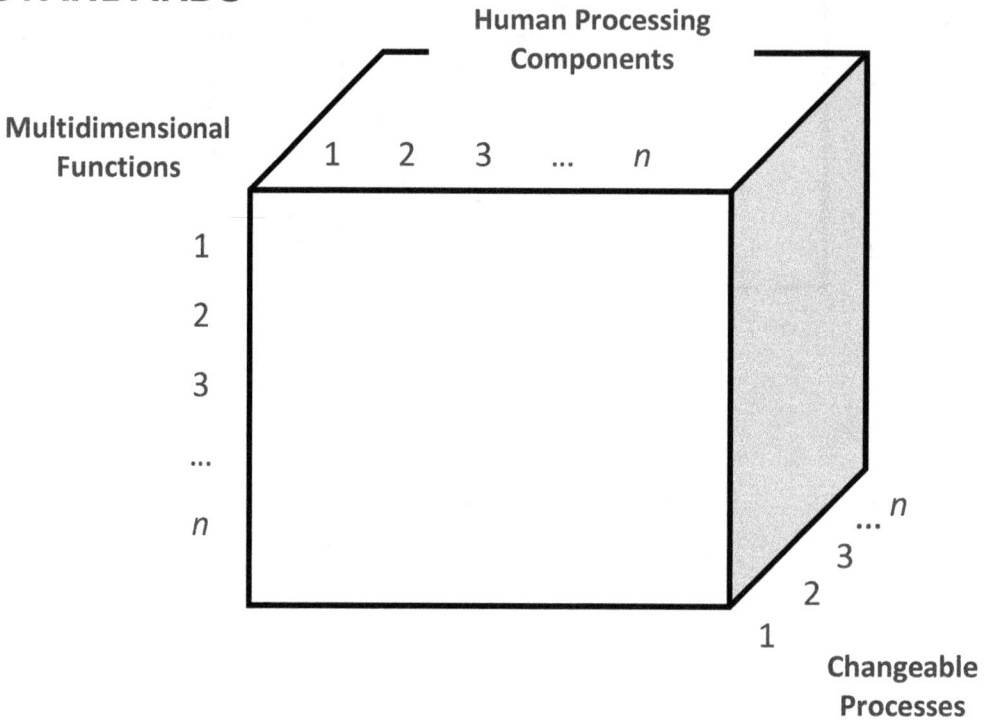

Figure 59. Operational Information Model for Standards

We may now view the "nested" models defining the phenomena (see Figure 60):

- **Generativity conditions**
- **Interdependent functions**
- **Multidimensional components**
- **Human processing processes**
- **Changeable standards**

We have defined the ideal dimensions for *unfettered thinking* or freedom.

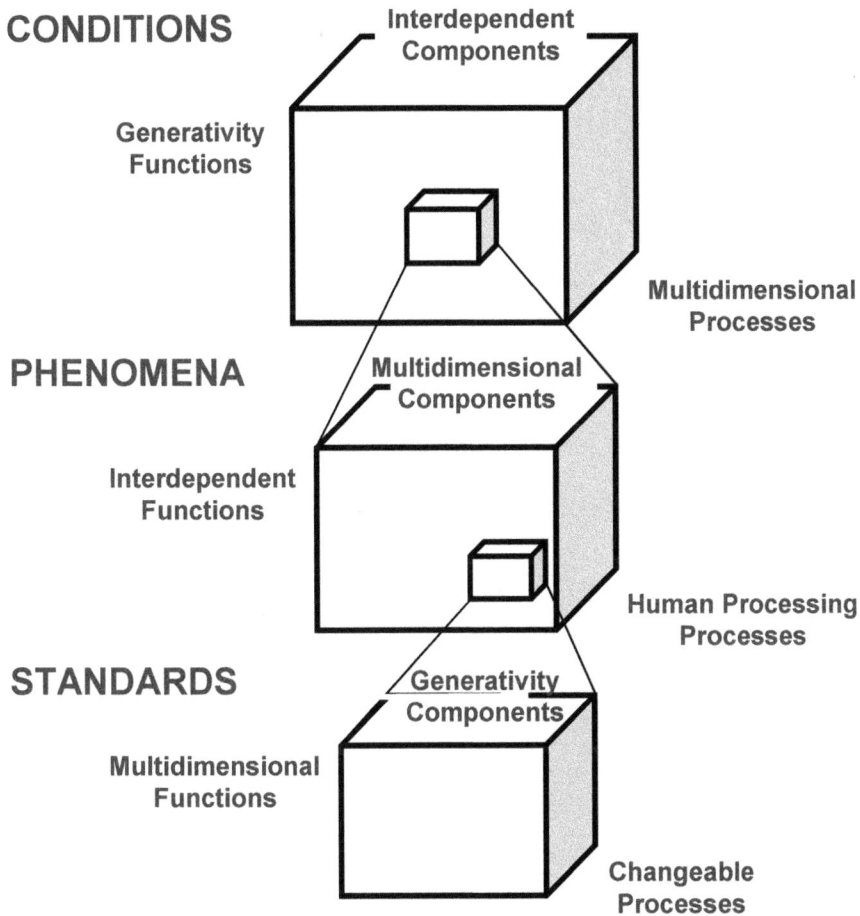

CONDITIONS

Interdependent
Components

Generativity
Functions

Multidimensional
Processes

PHENOMENA

Multidimensional
Components

Interdependent
Functions

Human Processing
Processes

STANDARDS

Generativity
Components

Multidimensional
Functions

Changeable
Processes

Figure 60. The Conditions, Functions, Components,
Processes, and Processes of Source Phenomena

9
"Something More..."

A century ago, "Nothing but..." scientists followed Einstein by projecting powerful force fields to be prepotent in the physical nature of the universe: mass, electric charge, weak, and strong nuclear charges.

Now Kakovitch (2012) discovers that during the Big Bang and shortly thereafter during the Radiation Era, sub-elementary particles and charges had not yet formed. Kakovitch asks and answers the basic questions of the nature of force:

> **What was the physical nature of force prior to the presence of matter which fixed the geometry of space-time? Could a scalar force-field exist that is solely a function of temperature? Can such a force-field relate to today's matter-dominated era?**

"Nothing but Platforms"

Let us illustrate with a model for **OCD** or **Organizational Capital Development**. Processing inductively, we have defined a multidimensional model for accomplishing scaled functions (see Figure 61):

- **Organizational functions**
- **Organizational components**
- **Organizational processes**

We may summarize these orthogonal **OCD** dimensions as follows:

> **Policy-driven functions are achieved by leadership-driven components empowered by goal-driven processing systems.**

In a healthy organization, we may say simply:

> **"We get what we process for!"**

OCD

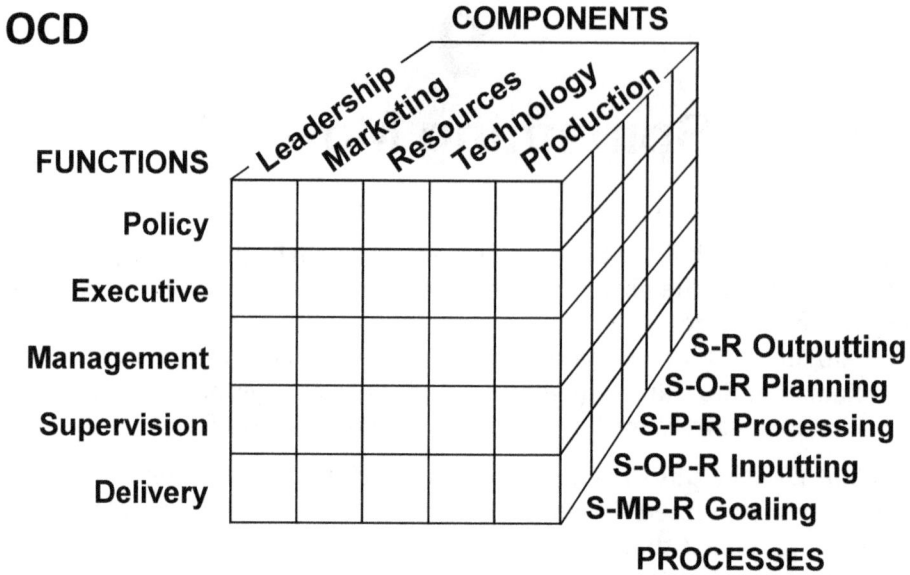

Figure 61. Modeling the Dimensions of OCD—
Organizational Capital Development

To continue the illustration, we may now seek out the Human Processing Source Cell or Processing Center of OCD. As may be noted, this Source Cell has its own scaled and orthogonal dimensions (see Figure 62):

- **Management functions**
- **Resource integration components**
- **Generative processing systems**

Once again, we may summarize the operations of the processing center:

> **Management functions are achieved by resource integration components empowered by generativity processes.**

In a healthy **"Processing Center,"** we may say simply:

"We generate a thinking organization."

OCD Processing Center

Components

Resource Integration

Technology

Production

Functions

Management

Supervision

Delivery

Outputting

Planning

Processing

Processes

OCD

Components

Leadership

Marketing

Resources

Technology

Production

Organizational Functions

Policy

Executive

Management

Supervision

Delivery

S-R Outputting

S-O-R Planning

S-P-R Processing

S-OP-R Inputting

S-MP-R Goaling

Organizational Processes

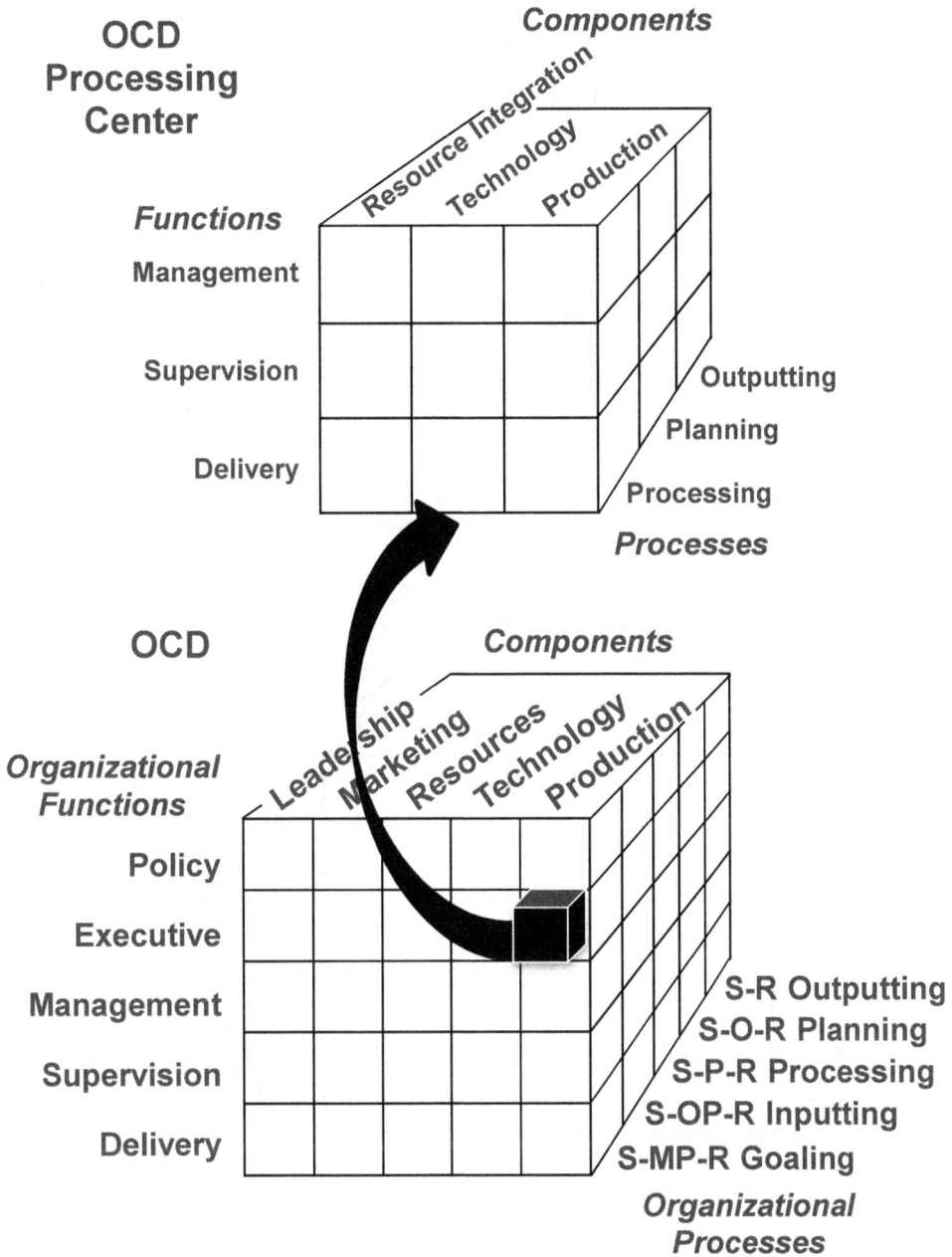

Figure 62. Modeling the "Source Cell" or "Processing Center"
of OCD—Organizational Capital Development

"Something More Hypothesis"

Arriving at the social schematics of our nothing but...platforms, we are now prepared to search for our something more...hypotheses.

As may be viewed, "Scientific Processors" may develop exploratory probes of the unknown spaces or conditions within which the processing center operates. Not unlike the space telescopes searching our universes, these initiatives are represented by balloons that carry us to greater heights and depths (see Figure 63).

As can be seen, these exploratory probes become nodes based upon the subject-matter expertise of the processors in the specialty dimension areas—functions, components, processes: F! C! P!

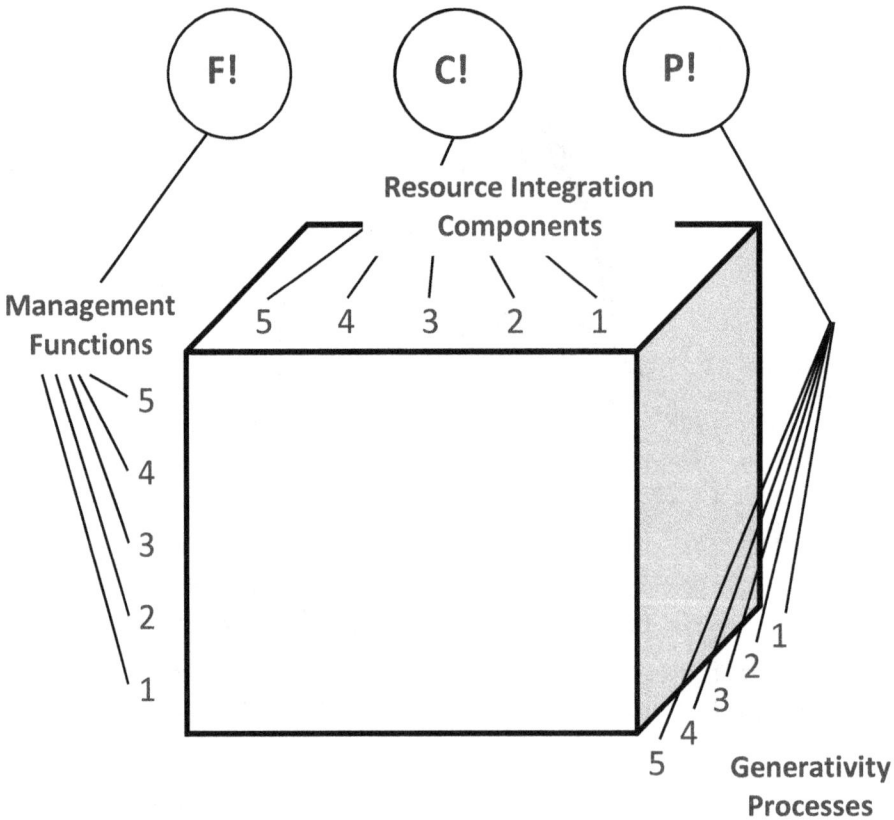

Figure 63. Modeling the "Processing Search" of the "Processing Center" of OCD (Deductive Model)

These interactions are indeed processing systems culminating in interdependent processing (see Figure 64).

As may be noted, the phases of processing occur as follows:

- **Individual thinking** processes (EUA) stimulus input (**S**) and generates new response outputs (**R'**).

- **Interpersonal collaboration** merges response outputs (R and R') from individual thinking into new responses (**R"**).

- **Interdependent processing** processes (EUA) interpersonal response output (**R"**) and generates most powerful new response (**R'''**).

Together, the individual, interpersonal, and interdependent processing systems constitute a *critical mass of generative processing.*

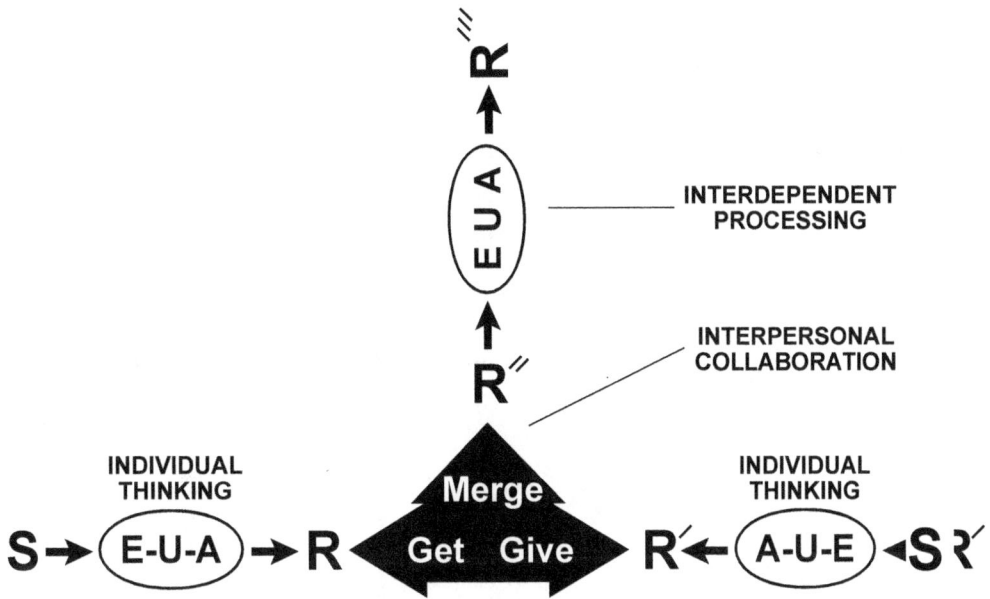

Figure 64. Modeling the Interdependent Processing of the Processing Center of OCD (Deductive Model)

Finally, we have generated our product, *marketplace positioning,* which details the significant discriminations to be made in positioning our unit or company for future business (see Figure 65):

- **Marketplace requirements** for targets where our values converge

- **Marketplace capacities** for what levels of resources empower us

- **Marketplace positioning** for where our intentionality directs us

Together, these functions, components, and processes operationally define the positioning within which our entity operates.

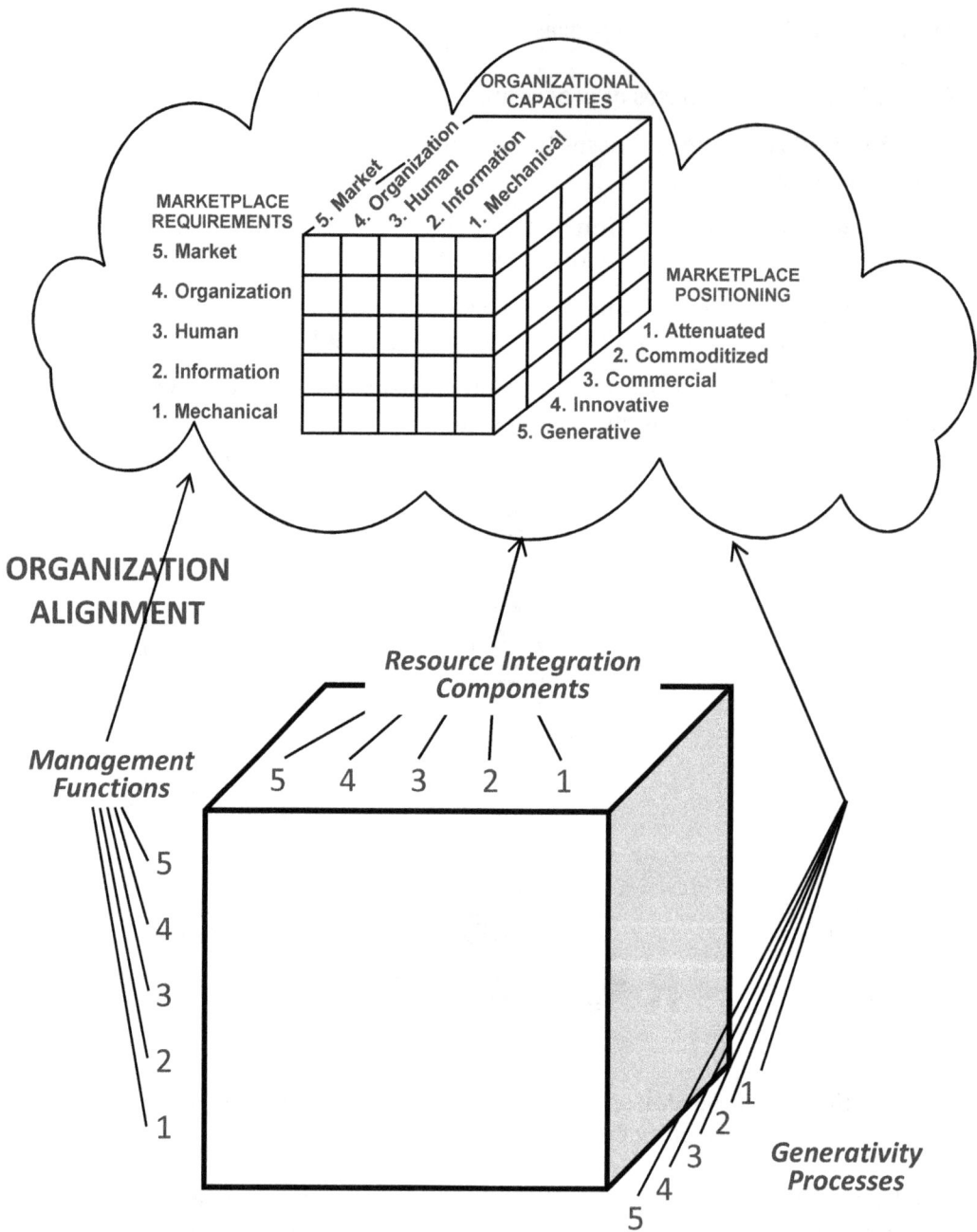

ORGANIZATION ALIGNMENT

Management Functions

Resource Integration Components

Generativity Processes

MARKETPLACE REQUIREMENTS

5. Market
4. Organization
3. Human
2. Information
1. Mechanical

ORGANIZATIONAL CAPACITIES

5. Market
4. Organization
3. Human
2. Information
1. Mechanical

MARKETPLACE POSITIONING

1. Attenuated
2. Commoditized
3. Commercial
4. Innovative
5. Generative

Figure 65. The Generative Product (MCD) of Interdependent Processing (Deductive Model)

To sum, we have discovered our future possibilities by building upon the **"Nothing More…"** of past probabilities. We have explored **"Something More…"** so that we could explain the space and define it. With a profound depth of commitment and follow-through, our explanation defines who we will become.

VI

Transition to the Future

10
Positioning for the Future—
A Case Study

So here we are. Our wisest, most hard-working, and most skilled account executives are near retirement. Our business has grown due to the qualitative efforts of these captains. They have built a sturdy ship of talented employees, charted a course that has kept us profitable for nearly 30 years, and serviced our client customers with products and services of high quality, at a low cost, and on time. But where does the company go from here? What will happen to our company once some of our most experienced and talented leaders retire? *Is this an impending crisis or an entrepreneurial opportunity?*

The obvious next step involves continuing to do what we have done in the past. This mean staying in the same business markets, servicing the same clients, and using the same skills and technologies we have been using over the last many years. They call this *sticking to knitting.* If you're good at knitting, stay in that business and just knit. Don't try anything else.

If we choose to *stick to knitting* we will still need to decide whether we want to grow or downsize.

Expanding our business in our current market would require us to focus our energies on the recruitment, selection, and training of replacements for those who will be retiring and for replacing those current employees who will be taking on higher-level responsibilities. Also necessary would be the expansion of our marketing function to bring in additional customers and additional contracts. Historically, our customers have asked us to work on many interesting projects. There is a certain amount of reinforcement employees experience when they have met the customers' challenges and delivered high-quality projects. We certainly could expand our employee base because our work is interesting. Like all jobs, there is still a certain level of "grind it out" work that simply must be done to satisfy a contract; yet, enough of the work requires empathic understanding of customer needs and the collection and representing of information that it is interesting. Business expansion by an investment of energies and resources into recruitment, selection, and training as well as marketing is a viable option.

If we decided to downsize by retirements and attrition, on the other hand, our gross would drop and we would not require a major effort or resources to recruit, select, and train new employees. In the downsizing scenario, with the retirement of some key account executives, we would also expect profitability levels to drop somewhat, per project and per person. Our contracts will continue to be interesting but we will have a smaller number of talented coworkers to bring to our table when solving customer problems. This too is a viable option, if that is what our remaining employees would like to do.

Beyond Downsizing

Due to the dynamic nature of our world, there is no way to escape requirements for generativity. Even if we are downsizing, all business options ultimately require generativity to enable growth or even survivability. The world changes, and we must change too. We can either change and grow in response to requirements that are imposed upon us by our customers, or we can impact future requirements with our own generative initiatives. Do we want to work in a future that others define, or do we want to apply our brainpower to contribute to defining the future for ourselves and others?

Living fully means living generatively. Yes, we would prefer to use our brainpower to help define the future. Yes, we want to contribute to a future of our own making. But, how can we possibly do it? For all of us, it begins with an energetic readiness, a motivated willingness, and the intellectual skills to generate new directions for ourselves and those we work with, and new directions for our customers as well. Generativity is a great opening of ourselves to possibilities that we have not yet even envisioned. The process of generating new directions is exciting. What if we could bring our customers breakthrough solutions? What if we had breakthroughs in how we do our own work that we could share with our customers? How exciting and motivational this would be for us and for our customers alike!

So where do we start? How do we enable generativity? At HTI we open ourselves to possibilities. We engage in a *get-give-merge* process of relating. We represent images of who we are and who we might become as clearly and comprehensively as we can. We draw systems diagrams, scales, matrices, and models to represent our ideas. We bring our reasoning skills to decide which ideas are essentially sound but would benefit from some strategic modifications. We expand options to go beyond our current representations to generate new ones. We narrow to choose new initiative courses of action, new hypotheses to test. We then prepare to bring our breakthrough initiatives to the marketplace. With project implementation, we will receive feedback that will serve as additional input for additional generative thinking!

Here are some of the exciting new ideas we have already generated but have not yet offered to our customer. These new breakthroughs may be used to reposition and grow HTI.

Breakthrough Ideation

We propose that the many intangeables in the marketplace are in fact discernable and definable. We label these ingredients New Capital, they are what is most important (see Table 5).

Table 5. NCD—New Capital Development

Marketplace capital – marketplace positioning

Organizational capital – organizational alignment

Human capital – human processing

Information capital – information modeling

Mechanical capital – mechanical tooling

Here are five critical sources of effectiveness. When you have efficient and effective machinery to service your needs, you have **Mechanical Capital**. The information you need is called **Information Capital**. A quality workforce is your **Human Capital**. An effective organization of distributed and coordinated efforts is **Organizational Capital**. The positioning of your organization and the products and services it delivers to the marketplace is your **Marketplace Capital**.

Interestingly, a corollary also holds true. The same five sources of effect needed by our customers are also needed by us, as producers. We also need high levels of marketplace, organizational, human, information, and mechanical capital. For producers, we title these same new capital ingredients slightly differently in that we call them *new capital technologies.* In other words, we need to know something about these new capital sources of effect because these new capital sources are the basis of the products or services we deliver. Fortunately, our decades of research and development have resulted in a unique understanding of these new capital sources of effect.

At HTI, we have developed the Customer Requirements and Producer Capabilities Matrix (see Table 6 on page 117) to help us understand the marketplace. This table enables us to map in our own current operations, including which technologies we currently rely on as we develop our products and services. We can also map in which sources of New Capital our customers are asking us to help them develop. As producers, we are extensions of our customer organizations. It is essential that we understand the needs of our customers so we can best service them. This matrix is one of many that we have generated to help us demystify the ingredients of effective performance for ourselves and for the customers we serve.

At HTI, we map ourselves into the Customer Requirements and Producer Capabilities Matrix as primarily a provider of information (Information Technologies column). We currently provide information to people who support human performance (the Human Capital row) within our customers' organizations. Some of our products also service people who work to develop or use information (knowledge workers in the Information Capital row). With this perspective in hand, we have gone on to analyze our current marketplace positioning and to consider how we would like to participate in the marketplace of products and services in the future. What follows is additional information about our new capital development technologies upon which we can position and invigorate HTI.

Table 6. Customer Requirements and Producer Capabilities

PRODUCER CAPABILITIES

CUSTOMER REQUIREMENTS	Marketplace Technologies	Organizational Technologies	Human Technologies	Information Technologies	Mechanical Technologies
Marketplace Capital	Marketplace segment that positions other marketplace segments	Organizational unit that assists in the position of marketplace segments (multiple organizations)	People who develop marketplace positioning (Policymakers)	Information that aids in positioning in the marketplace	Machines that aid in marketplace positioning
Organizational Capital	Marketplace segment that aligns organizational units	Organizational unit that assists in the alignment of organizational units	People who work to align organizational units (executives responsible for organizational realignment)	Information that assists in the alignment of organizational units	Machines used to assist in the alignment of organizations
Human Capital	Marketplace segment that supports human performance	Organizational unit that supports human performance (T&D, HRD Unit)	People who work to support human performance (manager, supervisors, trainers)	Information that assists and supports human performance	Machines used to assist and support human performance
Information Capital	Marketplace segment that supports the development and use of information	Organizational unit that supports the development and use of information (IT Unit)	People who work to support the development or use of information (knowledge workers)	Information that supports the development or use of information	Machines that support the development and use of information
Mechanical Capital	Marketplace segment that builds or operates machinery	Organizational unit that builds or operates machinery (Factory Operations Unit)	People who build or operate machines (machine-builders, machine operators)	Information that controls the building or operating of machinery	Machines for building machinery or other machine operations

Generativity in Understanding

If we have no unique contributions to make to our customers, then the longevity of our relationships with them will be limited. Once some other providers deliver their product or service either for a lower cost or in a more timely manner, in all likelihood, we will lose our customers unless we too drop back and compete in the commoditized and eventually attenuated market of cheaper and faster products and services. But what if we have unique, valued contributions we can make to our customers? What if we can do something better than anyone else? How can we generate to create breakthrough technologies? (By technologies we mean step-by-step methods or processes for accomplishing something.) How can we harness our generativity? This is the question that every organization and every individual needs to be asking to actualize marketplace as well as individual potential.

Here are some of the technological breakthroughs we have generated. It is delivered here in simple terms as an overview and sampler.

Marketplace Capital

This marketplace capital scale is one part of a 3D model for analyzing marketplace positioning (see Table 7). We can help organizations to see the positioning of their resources: machinery, information, people, and organizational units. With this same model, we can help policymakers see themselves within a larger marketplace of organizations, to see competitors as well as supplies. We are also able to see upstream and downstream for possible alliances and partnerships. Our marketplace capital technologies help us see how organizations can help others in the marketplace, how they might strategically team up for mutual benefits.

Our 3D marketplace perspective provides an insightful framework for policymaking. The technological models which follow add additional significant levels of valuable information useful for marketplace positioning (how markets, organizations, people, information, and products and services can strategically relate to maximize effectiveness.)

Table 7. Levels of Marketplace Capital

5. Generator
4. Innovator
3. Commercializer
2. Commoditizer
1. Attenuator

Organizational Capital

Our organizational technologies culminate in an operational virtual organization (see Table 8). This scale overviews how organizations are currently aligned and proposes new ways for units to work together.

Our own breakthroughs in organizational science impact Levels 3 through 5. We separate organizations by functional levels, component units, and common processes for accomplishing work. Once a 3D view of an organization is modeled, then significant new and informative organizational analyses can follow. With this information, many organizational alignment decisions can be more effectively made: internal resources can be strategically identified and deployed; specific information can be made available to the "right people at the right time"; commonalities across the organization can be seen and coordinated—"no need to reinvent the wheel."

Nested, 3D, and virtual organizations, at Levels 4 and 5 involve coordinating the efforts of suppliers and partners who outsource or extend the capabilities of the 3D organization.

We are excited about our organizational alignment approaches and a myriad of potential products and services that will flow from its implementation. We are excited too about the "effectiveness benefits" for customer organizations.

Table 8. Levels of Organizational Capital

5. Virtual Organization
4. Nested 3D Organization
3. 3D Organization
2. Matrixed Organization
1. Top Down Organization

Human Capital

Our Human Technology intervention is powerful (see Table 9). This scale focuses on Skills for human capital development—**The New 3Rs** or Relating, Representing, and Reasoning.

Human capital development has been a central focus of our research and projects over the last 30 years. At HTI, we have completed nearly 2000 Human Capital Development contracts. Our sister corporation, HRD Press has been a leading publisher of HRD or human resource development training materials for over 30 years as well. If a contribution to training and development has been made, we have been there as publishers, developers, and consultants to organizations. We have lived in the human capital development market for over 30 years.

Table 9. The New 3Rs—Human Capital

5. Reason by Acting
4. Reason by Narrowing
3. Reason by Expanding
2. Represent
1. Relate

Our HTI research and development unit has been working throughout these years and is prepared to deliver a marketplace breakthrough. With the growing requirement for thinking outside the box, **The New 3Rs** skills for generativity will be made available in the marketplace to elevate human performance.

We have researched the marketplace and there is no competition. We have the best, and perhaps only, true programmatic skills-set for generative thinking.

Information Capital

The next scale represents our information capital technologies (see Table 10). Most information representations found in the marketplace are verbal descriptions, databases of facts, and 2D spreadsheets. Yet, there are additional forms of information representation that can deliver better perspectives of the interactions of mission-critical information. It will be amazing how simplifying and demystifying information will become when more people begin to represent information using information capital approaches: information scales, nested scales, systems drawings, nested systems, 3D models, and even the multi-nesting of 3D models. We have developed new helpful approaches for representing information. We use these to stimulate generativity in ourselves, and they can be used for the same by our customers.

Table 10. Multi-D Information Capital

5. Multi-D, nested information
4. Nested 3D information
3. 3D information
2. 2D information
1. Linear information

What follows are recommendations for growing HTI. These recommendations are based upon the incorporation of breakthrough technologies for new capital development.

Generativity in Action

Is this a time of impending crisis or an entrepreneurial opportunity? With generative people we need not fear. Together, we'll come up with something new and better. It is the way of free, thinking people.

So, here we are, HTI, Human Technology, Inc. We are excited because we are poised with multiple new market-differentiated technologies. We have significant contributions to make to our customers' requirements for marketplace, organizational, human, and information capital. Because we have invested ourselves in generative thinking, we have an embarrassment of riches—"powerful ideation." Now, how can we transform our systematic technologies into products and services for our customers? What will we need to do to introduce these new capital technologies to our current and future customers?

Here is a simple listing of the action steps to transform the products or our generativity into action.

Human Technology, Inc.—Action Steps for Growth

1. Recruit, select, and train an HTI workforce to replace retiring account executives and to support the introduction of **new capital development** products and services.

2. Elevate our human capital performance offerings to feature skills for generative thinking—**The New 3Rs: Relating, Representing and Reasoning**.

3. Elevate our information capital development practices group offerings by applying the information representation technologies—**The 3Ss: Sentences, Systems and Schematics**.

4. Introduce a new organizational capital development group to deliver organizational consulting as well as products and services for aligning organizational resources: our breakthrough product/service—**CORe: Comprehensive Organizational Realignment.**

5. Introduce a new marketplace capital development group to deliver "multi-organization" consulting as well as deliver products and services with technologies for the positioning and relating of the resources of multiple-organizations: our breakthrough product/service—**Multi-CORe for Comprehensive Organizational Realignment** across **Multi**ple organizations.

This is an exciting, new, growthful direction for HTI based upon the products of our generative minds. None are directives from on high. They are offered for thoughtful consideration. What follows is a more detailed discussion of each of these new initiatives.

> 1. Recruit, select, and train an HTI workforce to replace retiring account executives and to support the introduction of **new capital development** products and services.

With highly valued account executives approaching retirement, consideration of elevating our offerings in Human Capital and Information Capital solutions, as well as the addition of organizational and marketplace new capital products and services, we will need to recruit, select and train account executives/consultants, project managers, and knowledge workers or information builders.

We need to consider **recruitment** strategies for finding high-potential hires. One pool may include soon-to-be graduating college seniors or graduate students. We will need to consider multiple avenues for finding talented employees.

A **selection** process that includes critical incident simulations could be of significant value. We would need to consider some form of a trial period for new employees who do not yet have a track record of employment.

Training will be a critical initiative. Our retiring executives have a deep knowledge of well-earned best practices. They are reservoirs of institutional knowledge of how our organization works and how our customer organizations work. New and elevated HTI employees who follow will be the beneficiaries of their accumulated wisdom. We will make every effort to encourage our retiring executives to share their knowledge while we support them to pursue their retirement activities.

In summary, recruitment, selection, and training are top priorities to the growth of HTI and will require financial and person-power resources to support them.

> 2. Elevate our human capital performance offerings to feature skills for generative thinking—**The New 3Rs: Relating, Representing and Reasoning**.

We begin our initiative thinking by modeling our potential marketplace initiatives. We revisit our Customer Requirements and Producer Capabilities Matrix (see Table 11). This chart shows one of the technological areas that is the core of our current products and services. We work primarily in the information technology cell highlighted in a light gray. The chart also displays the main customer requirements (highlighted in a medium gray) currently serviced by this unit. We are information product developers. We service people within our customers' organizations who manage, supervise, or otherwise support human performance effectiveness.

Table 11. Elevating Information Technology Products to Support Customer Human Performance Requirements

PRODUCER CAPABILITIES

CUSTOMER REQUIREMENTS	Marketplace Technologies	Organizational Technologies	Human Technologies	Information Technologies	Mechanical Technologies
Marketplace Capital	Marketplace segment that positions other marketplace segments	Organizational unit that assists in the position of marketplace segments (multiple organizations)	People who develop marketplace positioning (Policymakers)	Information that aids in positioning in the marketplace	Machines that aid in marketplace positioning
Organizational Capital	Marketplace segment that aligns organizational units	Organizational unit that assists in the alignment of organizational units	People who work to align organizational units (executives responsible for organizational realignment)	Information that assists in the alignment of organizational units	Machines used to assist in the alignment of organizations
Human Capital	Marketplace segment that supports human performance	Organizational unit that supports human performance (T&D, HRD Unit)	People who work to support human performance (manager, supervisors, trainers)	Information that assists and supports human performance	Machines used to assist and support human performance
Information Capital	Marketplace segment that supports the development and use of information	Organizational unit that supports the development and use of information (IT Unit)	People who work to support the development or use of information (knowledge workers)	Information that supports the development or use of information	Machines that support the development and use of information
Mechanical Capital	Marketplace segment that builds or operates machinery	Organizational unit that builds or operates machinery (Factory Operations Unit)	People who build or operate machines (machine builders, machine operators)	Information that controls the building or operating of machinery	Machines for building machinery or other machine operations

Now that we are mapped-into our marketplace view, we can perform a more thorough analysis of both our customer requirements and the technologies we employ to service these requirements. Are the customer requirements for human capital products and services changing? Will the technological approaches we are currently using be enough to satisfy our customers or will they be demanding some new and better approaches?

At HTI, we have developed nearly a thousand training courses to elevate human performance. The interpersonal and instructional design approaches we currently use to help organizations develop their human capital is tops in the industry. We have won many competitions to prove it and our customers keep coming to us to perform additional work. We have been doing much right. The marketplace, however, never stands still. It is evolving and we can follow or lead.

With the vast quantities of information now being produced, collected, and made available, we, and our customers, are becoming overloaded. The once simple is now extremely complex. Customer organizations have a growing need to empower employees to factor, demystify, and relate information—discover its meaning and value and then use it effectively or pass it on to someone else who needs it.

Customers also have a growing need for people who can go beyond collected information to create new responses, products, and services to responsively solve seemingly intractable problems or initiatively take advantage of marketplace opportunities. Generative thinking is the engine of entrepreneurial economics and the source of solutions in both the public and private sectors. Succinctly, the evolving requirement for human capital is higher order skills for generative thinking.

We are excited to have developed a next generation skills technology for human performance: **The New 3Rs of Relating, Representing and Reasoning**. It is fully developed and tested. We have yet-unpublished training materials prepared. **The New 3Rs** are effective and face-valid. These are skills that make sense to people once they are introduced to them.

These are new core skills for human capital development. **The New 3Rs** are a powerfully effective skills package as they enable generativity. These same **New 3Rs** skills are also extremely efficient for our customers as basic interpersonal, informational, problem-solving, creativity, goal-setting, program development, decision-making, negotiation, teamwork, mentoring, supervisor skills, and many more areas of human performance are enabled by these core skills. The majority of the hundred or more specific human "soft skills" courses currently offered by organizations will now be best serviced by training in **The New 3Rs** as a core skills-set with shorter training modules added to explain and explore problem-specific terms, policies, and laws. (As an example, the skills requirements for contributing to a team effort are most efficiently and effectively addressed by empowerment in **The New 3Rs** with a limited amount of information and additional skills that are specific to teamwork in a specific organization.) This means that **The New 3Rs** delivers more for less—more effective and less costly.

The New 3Rs of Relating, Representing and Reasoning is a technological breakthrough and the basis for a new focus by our customers' Training and Development (T&D) units. T&D units are in need of a more effective and efficient approach to Human Capital Development. The solution is employee empowerment in **The New 3Rs.**

3. Elevate our Information Capital Development practices group offerings by applying the information representation technologies—**The 3Ss: Sentences, Systems and Schematics**.

HTI is also an information developer that services our customer organizations' information capital development needs. This means that we develop information that helps people develop or use information more effectively (see Table 12). The information products we develop are represented by the technology cell highlighted in light gray. This unit services our customers' requirements (highlighted in medium gray).

Over 25 years ago we, at HTI, developed breakthrough texts in Instructional Systems Design or ISD. This process introduced a series of systematic methods for analyzing information. We built information for our customers: from facts to concepts to principles to applications to programs. We analyzed our customers' contextual requirements: goals through tasks and skills and finally to skills steps. We have performed job task analyses for hundreds of organizations and many more hundreds of additional information analyses for our customer organizations. Our projects have always involved factoring and demystifying information. We have used our information-building approaches with good results. Our customers have found our information products to be of the highest caliber.

Along the way we have had to also conquer many information delivery technologies: publishing, database design and development, video, online interactive programs, web design, and more. We have performed admirably, satisfying customer requirements in this market space. We have reason to be proud of our accomplishments and proud of our people who have conquered their information-building substance and information delivery technologies.

New technologies for delivering information to users continues to change at a rapid pace. Like every information development company, we expend considerable effort to keep abreast of new delivery software and hardware. We do a good job using available information mediums however we follow what is available from the market. We do not yet claim any market differentiation in our use of software or hardware to deliver information.

Today, we are no longer unique in the marketplace of information developers as we once were when we initially introduced our Instructional Design Technologies. This need not remain the case however. We have developed new information representation technologies (step-by-step processes) for organizing, analyzing and presenting information. We call our information systems representation technology **The 3Ss— Sentences, Systems and Schematics**.

Table 12. Elevating Information Technology Products to Support the Development and Use of Information

PRODUCER CAPABILITIES

CUSTOMER REQUIREMENTS	Marketplace Technologies	Organizational Technologies	Human Technologies	Information Technologies	Mechanical Technologies
Marketplace Capital	Marketplace segment that positions other marketplace segments	Organizational unit that assists in the position of marketplace segments (multiple organizations)	People who develop marketplace positioning (Policymakers)	Information that aids in positioning in the marketplace	Machines that aid in marketplace positioning
Organizational Capital	Marketplace segment that aligns organizational units	Organizational unit that assists in the alignment of organizational units	People who work to align organizational units (executives responsible for organizational realignment)	Information that assists in the alignment of organizational units	Machines used to assist in the alignment of organizations
Human Capital	Marketplace segment that supports human performance	Organizational unit that supports human performance (T&D, HRD Unit)	People who work to support human performance (manager, supervisors, trainers)	Information that assists and supports human performance	Machines used to assist and support human performance
Information Capital	Marketplace segment that supports the development and use of information	Organizational unit that supports the development and use of information (IT Unit)	People who work to support the development or use of information (knowledge workers)	Information that supports the development or use of information	Machines that support the development and use of information
Mechanical Capital	Marketplace segment that builds or operates machinery	Organizational unit that builds or operates machinery (Factory Operations Unit)	People who build or operate machines (machine-builders, machine operators)	Information that controls the building or operating of machinery	Machines for building machinery or other machine operations

113

The 3Ss strategies for information representation call for comprehensiveness in verbal communications and expanded use of systems diagrams. What makes us different, however, and more effective than other information developers are our technologies for representing information schematically. We have developed useful methodologies for scaling, matrixing, 3D modeling, nested-modeling, and multi-nested 3D modeling. These representational strategies enable users to build information representations that offer elevated levels of strategic perspective. Seeing how constructs (ideas represented with words) relate to other constructs, within schematic (visual) models, delivers unique and useful perspectives. A picture truly is often worth a thousand words. **The 3Ss**, not yet published or released to the general public, are potent technologies for information capital development. We can use **The 3Ss** in the development of products for our customers and also empower our customers to use it as well. This is yet another breakthrough technology that we use to grow our market presence helping our customers, develop and use information.

---

4.   Introduce a new organizational capital development group to deliver organizational consulting as well as products and services for aligning organizational resources: our breakthrough product/service—**CORe: Comprehensive Organizational Realignment.**

Organizational capital is defined as the alignment of the resources of an organization. This means the alignment of organizational units, people, information, machinery, equipment, and goods. The leaders of our government agencies have explicitly stated that resource alignment and information sharing are strategic organizational goals. Organizational resource alignment is a critical goal of private sector organizations as well.

We have developed new technologies to service an organization's resource alignment needs (see Table 13). We have developed a 3D organizational model, a nested-3D model, and a virtual organizational model to represent the design of the, much-sought-after, interdependent organization with strategically coordinated efforts and leveraged, time-sensitive information sharing. The 3D organization and its higher-level variations can be offered to replace ineffective, bottle-necked, top-down organizational structures as well as confused, matrix-management organizations.

We can offer organizational consulting analyses and reports or training for executives or executive teams responsible for organizational change initiatives. We can also provide information products—designs for servicing the alignment of information within and between organizational units. Our organizational models enable recommendations for the direction and flow of vital information within and between organizations. We have developed preliminary designs for a unique database software system that would support the identification, collection, and distribution of critical and timely information across an organization. We call our organizational design CORe—Comprehensive Organizational Realignment.

Our CORe initiative may begin as an organizational consulting contract, a policymaker training intervention, or an information product (anywhere from paper design through software development).

Table 13. Servicing Customers' Organizational Requirements with Organizational Technologies

PRODUCER CAPABILITIES

CUSTOMER REQUIREMENTS	Marketplace Technologies	Organizational Technologies	Human Technologies	Information Technologies	Mechanical Technologies
Marketplace Capital	Marketplace segment that positions other marketplace segments	Organizational unit that assists in the position of marketplace segments (multiple organizations)	People who develop marketplace positioning (Policymakers)	Information that aids in positioning in the marketplace	Machines that aid in marketplace positioning
Organizational Capital	Marketplace segment that aligns organizational units	Organizational unit that assists in the alignment of organizational units	People who work to align organizational units (executives responsible for organizational realignment)	Information that assists in the alignment of organizational units	Machines used to assist in the alignment of organizations
Human Capital	Marketplace segment that supports human performance	Organizational unit that supports human performance (T&D, HRD Unit)	People who work to support human performance (manager, supervisors, trainers)	Information that assists and supports human performance	Machines used to assist and support human performance
Information Capital	Marketplace segment that supports the development and use of information	Organizational unit that supports the development and use of information (IT Unit)	People who work to support the development or use of information (knowledge workers)	Information that supports the development or use of information	Machines that support the development and use of information
Mechanical Capital	Marketplace segment that builds or operates machinery	Organizational unit that builds or operates machinery (Factory Operations Unit)	People who build or operate machines (machine builders, machine operators)	Information that controls the building or operating of machinery	Machines for building machinery or other machine operations

CORe incorporates a 3D view of an organization and includes representations of the nesting of resources and processes within the organization. CORe educates users so they may see how they contribute to larger organizational goals. They can see how the many goals of their organizations align in the service or a larger mission and how their activities and results impact lower-level activities and results. CORe also enables the identification, relational-mapping, tracking, and distribution of information across an organization. The goal of CORe is to support the alignment of the organization's people and material resources, activities, and information. CORe will contribute to increased performance effectiveness and efficiency (the right information, to the right people, and as instantaneously as it is made available.)

Additionally, any organizational alignment initiative brings to light every organizations' need to elevate human performance (**The New 3Rs**) and its need to develop and disseminate more valued information (**The 3Ss**).

---

5. Introduce a new marketplace capital development group to deliver "multi-organization" consulting as well as deliver products and services with technologies for the positioning and relating of the resources of multiple-organizations: our breakthrough product/service—**Multi-CORe** for **Comprehensive Organizational Realignment** across **Multi**ple organizations.

No organization is autonomous or independent. All organizations exist in relation to other organizations. They may share similar goals, offer similar products or services, or service the same customers. Each organization impacts others and is impacted by others. This results in coordination problems and partnership opportunities between and among organizations. How can organizations manage these complex multiorganization relationships?

We have developed a solution to multiorganization problems and opportunities. We call our approach **Multi-CORe** for **Comprehensive Organizational Realignment** across **Multi**ple organizations.

Multi-CORe helps organizations identify opportunities for relationship building. Multi-CORe supports the representation of an organization, its initiatives, products, services, or any other ingredient it identifies as important, in 1D, 2D, 3D, or multiple nested 3D maps. It also supports the identification and representation of other organizations using the same methods and then integrates this information. This results in visual representation of relationships among organizations. Users can see who and what is sharing the same market space, who and what is upstream (affecting it), who and what is downstream (being affected by it). These perspectives generate opportunities to identify future problems before they become reality or to identify them and solve them before they grow larger.

The Multi-CORe approach for the positioning of multiple organizations is similar to CORe for alignment within individual organizations. An intervention can begin as a consulting experience with a group of policymakers from multiple organizations who are tasked with cross-organizational alignment. It can begin as a training experience for the policymakers of a single organization who want to maximize the marketplace positioning of their organization and their products (partnerships, mergers and acquisitions). Or it may begin with interest in an information product to assist an organization with its positioning/alignment in the marketplace of multiple organizations. HTI, with possession of the Multi-CORe technogy for marketplace position, can deliver these services and products (see Table 14).

The requirement for coordination among multiple organizations is here, and here to stay.

Additionally, any multiorganization marketplace positioning initiative brings to light each individual organizations' need to improve organizational alignment (**CORe**), elevate human performance (**The New 3Rs**), and its need to develop and disseminate more valued information (**The 3Ss**).

Table 14. Servicing Requirements for Coordination among Multiple Organizations with Marketplace Technologies

PRODUCER CAPABILITIES

CUSTOMER REQUIREMENTS	Marketplace Technologies	Organizational Technologies	Human Technologies	Information Technologies	Mechanical Technologies
Marketplace Capital	Marketplace segment that positions other marketplace segments	Organizational unit that assists in the position of marketplace segments (multiple organizations)	People who develop marketplace positioning (Policymakers)	Information that aids in positioning in the marketplace	Machines that aid in marketplace positioning
Organizational Capital	Marketplace segment that aligns organizational units	Organizational unit that assists in the alignment of organizational units	People who work to align organizational units (executives responsible for organizational realignment)	Information that assists in the alignment of organizational units	Machines used to assist in the alignment of organizations
Human Capital	Marketplace segment that supports human performance	Organizational unit that supports human performance (T&D, HRD Unit)	People who work to support human performance (manager, supervisors, trainers)	Information that assists and supports human performance	Machines used to assist and support human performance
Information Capital	Marketplace segment that supports the development and use of information	Organizational unit that supports the development and use of information (IT Unit)	People who work to support the development or use of information (knowledge workers)	Information that supports the development or use of information	Machines that support the development and use of information
Mechanical Capital	Marketplace segment that builds or operates machinery	Organizational unit that builds or operates machinery (Factory Operations Unit)	People who build or operate machines (machine builders, machine operators)	Information that controls the building or operating of machinery	Machines for building machinery or other machine operations

Summary and Transition

When we founded HTI in 1980, we had already spent the better part of two years surveying more than 100 government agencies in order to discriminate the most urgent need that we could meet. When we met Col. Everett at the Office of Personnel Management (OPM) and he expressed instructional design as the unfilled space, we answered, "We wrote the book!" Three months later, we handed him a draft of *Instructional Systems Design.* Shortly thereafter, we formally published our first book on ISD and co-processed with OPM to generate the Indefinite Quantity Contracting Vehicle to enable government agencies to contract for projects with pre-qualified, product development contractor companies.

Now we find ourselves in a similar, if not the identically-same, situation as in the early 1980s. After thousands of projects and dramatically changing conditions such as natural disasters (Katrina) and human catastrophes (9/11) and the continuing failure of our government agencies to respond effectively, we are now able to discriminate our customers' new, most urgent unfilled needs: organizational alignment and alignment among multiple organizations. Now we can answer, "We have the step-by-step technology!" Shortly, we will be ready to show our customers our technological approaches and be ready to empower them and co-process with them.

We are pleased to have developed our new capital technologies: marketplace, organization, human, and information. We are excited to have developed commercial, step-by-step, technological designs: **Multi-CORe, CORe, The New 3Rs, and The 3Ss.**

Here is a single page visual summary of a growth strategy for HTI, Human Technology, Inc. (see Table 15).

The black cells show opportunities for HTI to service marketplace positioning requirements—the need for organizations to position and strategically partner with other organizations. Our technological offering is **Multi-CORe—Comprehensive Organizational Realignment among Multiple Organizations.**

The dark gray cells present business opportunities for HTI to assist customer organizations with the alignment of their resources: units, people, information, and capital goods. Our technological response is **CORe—Comprehensive Organizational Realignment.**

The medium gray cells highlight the customer requirements that our current technologies (light gray cells) service. At the Human Capital requirements level (center row), we can reposition our human performance product and service offerings with a new focus on skills for thinking and generativity. We can incorporate **The New 3Rs—Relating, Representing and Reasoning** as the leveraged Human Capital skills-set for T&D and HRD. At the Information Capital requirements level (second row from the bottom) we can reposition our information products with an emphasis upon our ability to incorporate elevated forms of information representation, in particular, schematic representations. We call our technology **The 3Ss—Sentences, Systems and Schematics.**

Together, these four new technological offerings are exciting to consider. We are once again on the cutting edge. We can bring new solutions to pressing customer problems and opportunities.

Table 15. Elevating the Product and Service Offerings of Human Technology, Inc.

PRODUCER CAPABILITIES

CUSTOMER REQUIREMENTS	Marketplace Technologies	Organizational Technologies	Human Technologies	Information Technologies	Mechanical Technologies
Marketplace Capital	Marketplace segment that positions other marketplace segments	Organizational unit that assists in the position of marketplace segments (multiple organizations)	People who develop marketplace positioning (Policymakers)	Information that aids in positioning in the marketplace	Machines that aid in marketplace positioning
Organizational Capital	Marketplace segment that aligns organizational units	Organizational unit that assists in the alignment of organizational units	People who work to align organizational units (executives responsible for organizational realignment)	Information that assists in the alignment of organizational units	Machines used to assist in the alignment of organizations
Human Capital	Marketplace segment that supports human performance	Organizational unit that supports human performance (T&D, HRD Unit)	People who work to support human performance (manager, supervisors, trainers)	Information that assists and supports human performance	Machines used to assist and support human performance
Information Capital	Marketplace segment that supports the development and use of information	Organizational unit that supports the development and use of information (IT Unit)	People who work to support the development or use of information (knowledge workers)	Information that supports the development or use of information	Machines that support the development and use of information
Mechanical Capital	Marketplace segment that builds or operates machinery	Organizational unit that builds or operates machinery (Factory Operations Unit)	People who build or operate machines (machine builders, machine operators)	Information that controls the building or operating of machinery	Machines for building machinery or other machine operations

We are excited about the future of HTI. We wish our older and soon-to-be retiring account executives will stay a little longer and join us for the upcoming marketplace introduction of our new capital technologies. We are truly grateful for the human capital that they are and the dedication, support, and leadership they have provided over the past many years.

In transition, this report is offered as stimulus for further thinking and consideration by the generative minds who together form Human Technology, Inc. You have paid the price of honest empathy with your customers and worked countless hours. You have conquered content and technologies to become qualitative information builders. Empathic initiative requires incredible stamina, and we must attend to replenishing our physical resources. You deserve some R&R (yes, rest and relaxation), there is no doubt. Let's figure out a way to get you some!

Then, let's consider the future of HTI. It is exhilarating and satisfying to live on the cutting edge of our information-driven world. And now, we have new capital technologies to empower people to solve seemingly intractable problems and to generate new entrepreneurial initiatives. It is an exciting time.

We define ourselves by the hypotheses we will decide to test!

What qualifies HTI to respond to this need and design such a system?! Simply this: HTI has been the exemplary performer in the OPM vehicle over more than one quarter century! The experience and expertise of HTI executives and managers demonstrate the capacity for formulating organizational DNA designs (discriminating objects and the levels of scales within them). Moreover, HTI executives have the capacity to co-process with its R&D in Carkhuff Thinking Systems, which has already piloted **CORE**.

Perhaps the greatest and most enduring contribution that HTI can make is to design the **CORE** that communicates the presence and movement of all assets and conditions within, between, and among all agencies. The **CORE** would empower leaders to track all assets under changing condition, thus enabling them to align assets to respond effectively to crises. The implications of transfer to the private sector are ubiquitous.

11
"God's Work"

In transition, the culminating implications of probabilities and possibilities sciences contrast vividly: infinitesimal to infinity! Given the constancy of linear, independent, and symmetrical phenomena, probabilities science is planning-centric. In contrast, given the changeability of multidimensional, interdependent, and asymmetrical phenomena, possibilities science is process-centric. Only continuous processing can align, empower, and free the perpetual motion of changing phenomena.

The fundamental principles of possibilities science thus empower us to generate infinite possibilities that both embrace and resolve the apparent conflicts of relativity and chaos: multidimensionality, interdependence, asymmetrical curvilinearity, and changeability. We can neither explore our universes nor penetrate infinity with the simplicity of the equations of probabilities science: $x = y$.

Indeed, the models of possibilities science that we have represented are highly initiative and, as such, have limitations even as they move inexorably toward discovering the nature of nature and its phenomena: multidimensional, interdependent, asymmetrical, changeable. This is **human-kind's work.**

The phenomenal models, in turn, require a higher level of alignment through relating; of enhancing phenomenal potential through empowerment; of freeing phenomena to discover their own changeable destinies. They are therefore experiential, evolving, and unifying; they express the changing basic fabric of developmental merging toward singularity and the resulting explosion of infinite possibilities. This is **God's Work.**

References

- Information Sciences
- Human Sciences

Information Sciences

Avery, J. *Information Theory and Evolution.* Singapore: World Scientific, 2003.

Bairstow, J. *The Father of the Information Age.* New York: Laser Focus World, 2002.

Bateson, G. *Form, Substance, and Difference in Steps to an Ecology of the Mind.* Chicago, IL: University of Chicago, 1972.

Bekenstein, J. D. Information in the Holographic Universe. *Scientific American,* 2003.

Beynon-Davies, P. *Information Systems: An Introduction to Informatics in Organization.* London: Palgrave, 2002.

Beynon-Davies, P. *Business Information Systems.* London: Palgrave, 2009.

Brown, J. S. and Duguid, P. *The Social Life of Information.* Boston, MA: Harvard Business School, 2002.

Casagrande, D. Information as Verb: Re-conceptualizing Information for Cognitive and Ecological Models. *Journal of Ecological Anthropology,* 1999, 3, pp. 4–13.

Dusenbery, D. B. *Sensory Ecology.* New York: Q. R. Freeman, 1992.

Floridi, L. *Information: A Very Short Introduction.* Oxford: Oxford University Press, 2010.

Gleick, J. *What Just Happened: A Chronicle from the Information Frontier.* New York: Pantheon, 2002.

Gleick, J. *The Information: A History, a Theory, a Flood.* New York: Pantheon, 2011.

Goonatilake, S. *The Evolution of Information.* London: Pinter, 1991.

Headrick, D. R. *When Information Came of Age.* Oxford: Oxford University Press, 2000.

Hoagland, M. and Dodson, B. *The Way Life Works.* New York: Random House, 1995.

Liu, A. *The Laws of Cool: Knowledge, Work and the Culture of Information.* Chicago, IL: University of Chicago, 2004.

Pérez-Montoro, M. *The Phenomenon of Information.* Lanham, MD: Scarecrow, 2007.

Roederer, J. G. *Information and Its Role in Nature.* Berlin: Springor, 2005.

Seife, C. *Decoding the Universe.* New York: Viking, 2006.

Solymar, L. *Getting the Message: A History of Communication.* Oxford: Oxford University Press, 1999.

Stewart, T. *Wealth of Knowledge.* New York: Doubleday, 2001.

Vigo, R. Representational Information. *Information Sciences.* 2011, 181, 4847–4859.

Virilio, P. *The Information Bomb.* London: Verso, 2000.

Watson, J. *Behaviorism.* Chicago, IL: University of Chicago Press, 1930.

Wicker, S. B. and Kim, S. *Fundamentals of Codes, Graphs, and Iterative Decoding.* New York: Sprager, 2003.

Yockey, H. P. *Information Theory and the Origins of Life.* Cambridge: Cambridge University Press, 2005.

Young, P. *The Nature of Information.* Westport, CT: Greenwood, 1987.

Human Sciences

Anthony, W. *The Principles of Psychiatric Rehabilitation.* Baltimore, MD: University Park Press, 1979.

Aspy, D. N., and Roebuck, F. N. *Kids Don't Learn from People They Don't Like.* Amherst, MA: HRD Press, 1977.

Banks, G. P. *From Bondage through Prosperity: Finding the Freedom in Thinking.* Amherst, MA: HRD Press, 2013.

Berenson, B. G. *The Possibilities Mind.* Amherst, MA: HRD Press, 2001.

Berenson, B. G. *Carkhuff and The Human Sciences.* McLean, VA: The McLean Project, 2013.

Berenson, B. G. and Cannon, J. R. *The Science of Freedom.* Amherst, MA: HRD Press, 2006.

Berners-Lee, T. *Weaving the Web.* Britain: Orion Business, 1989, ISBN 0-7528-2090-7.

Bierman, R. *Toward Meeting Fundamental Human Service Needs.* Guelph, Ontario: Human Service Community, Inc., 1976.

Bugelski, B. R. *Psychology of Learning.* New York: Holt, Rinehart & Winston, 1956.

Bugelski, B. R. *Principles of Learning.* New York: Praeger, 1979.

Carkhuff, R. R. *Helping and Human Relations. Volumes I and II.* New York: Holt, Rinehart & Winston, 1969.

Carkhuff, R. R. *The Development of Human Resources.* New York: Holt, Rinehart & Winston, 1971.

Carkhuff, R. R. *The Promise of America.* Amherst, MA: HRD Press, 1974.

Carkhuff, R. R. *Toward Actualizing Human Potential.* Amherst, MA: HRD Press, 1981.

Carkhuff, R. R. *Sources of Human Productivity.* Amherst, MA: HRD Press, 1983.

Carkhuff, R. R. *The Exemplar.* Amherst, MA: HRD Press, 1984.

Carkhuff, R. R. *Human Processing and Human Productivity.* Amherst, MA: HRD Press, 1986.

Carkhuff, R. R. *The Age of the New Capitalism.* Amherst, MA: HRD Press, 1988.

Carkhuff, R. R. *Empowering.* Amherst, MA: HRD Press, 1989.

Carkhuff, R. R. *Human Possibilities.* Amherst, MA: HRD Press, 2000.

Carkhuff, R. R. *The Age of Ideation.* Amherst, MA: HRD Press, 2007.

Carkhuff, R. R. *The Art of Helping.* Ninth Edition. Amherst, MA: HRD Press, 2009.

Carkhuff, R. R. *Saving America: The Generativity Solution.* Amherst, MA: HRD Press, 2010.

Carkhuff, R. R. *TheMcLeanProject.com.* The McLean Project, 2011.

Carkhuff, R. R. *Human Generativity: An Introduction to Human Sciences.* Amherst, MA: HRD Press, 2013.

Carkhuff, R.R. *The Human Sciences: Volume I. Probabilities, Possibilities, and Generativity Sciences.* Amherst, MA: HRD Press, 2013.

Carkhuff, R.R. *The Human Sciences: Volume II. Probabilities, Possibilities, and Generativity Technologies.* Amherst, MA: HRD Press, 2013.

Carkhuff, R. R. and Berenson, B. G. *The New Science of Possibilities. Volumes I and II.* Amherst, MA: HRD Press, 2000.

Carkhuff, R. R. and Berenson, B. G., et al. *The Possibilities Leader.* Amherst, MA: HRD Press, 2000.

Carkhuff, R. R. and Berenson, B. G., et al. *The Possibilities Organization.* Amherst, MA: HRD Press, 2000.

Carkhuff, R. R. and Berenson, B. G., et al. *Freedom-Building.* Amherst, MA: HRD Press, 2003.

Carkhuff, R. R. and Berenson, B. G., et al. *The Freedom Doctrine.* Amherst, MA: HRD Press, 2003.

Carkhuff, R. R. and Berenson, B. G., et al. *The Freedom Wars.* Amherst, MA: HRD Press, 2005.

Carkhuff, R. R. and Berenson, B. G., et al. *The Possibilities Economy.* Amherst, MA: HRD Press, 2006.

Carnavale, A. *Human Capital.* Washington, DC: American Society for Training and Development, 1983.

Drasgow, J. Eclipsing All Great Works. Foreword, *The Freedom Wars.* Amherst, MA: HRD Press, 2000.

Einstein, A. *Relativity: The Special and General Theory.* New York: Henry Holt, 1931.

Einstein, A. *The Evolution of Physics.* Cambridge: University of Cambridge, 1938.

Einstein, A. *Collected Papers of Albert Einstein.* Princeton, NJ: Princeton University Press, 1989.

References

Hebb, D. O. *The Organization of Behavior.* New York: John Wiley and Sons, 1949.

Hull, C. L. *Mathematics—Deductive Theory of Rote Learning.* New York: Appleton–Century–Crofts, 1940.

Hull, C. L. *Principles of Behavior.* New York: Appleton–Century–Crofts, 1943.

Hull, C. L. *A Behavior System.* New Haven, CT: Yale University Press, 1952.

Kakovitch, T. *Collegium.* Amherst, MA: HRD Press, 2012.

Kakovitch, T. *The Fifth Force.* Amherst, MA: HRD Press, 2012.

Kakovitch, T. *Anthropogenics.* Amherst, MA: HRD Press, 2013.

Kilby, J. *First Successful Demonstration of Integrating a Transition with Resistors and Capacitors on a Simple Semiconductor Chip Defining the Monolithic Idea.* Dallas, TX: Texas Instruments, September 12, 1958.

Pavlov, I. P. *Conditioned Reflexes.* Oxford: Oxford University Press, 1927.

Rogers, C. R. The Necessary and Sufficient Conditions of Therapeutic Personality Change. *Journal of Consulting Psychology,* 1957, 22, 95–103.

Siegel, S. *Nonparametric Statistics for the Behavioral Sciences.* Washington, DC: American Association for the Advancement of Science, 1959.

Sprinthall, R. C. *SPSS.* Boston, MA: Pearson Education, Inc., 2009.

Sprinthall, R. C. *Psychenomics.* Afterword in R. R. Carkhuff, *Saving America: The Generativity Solution.* Amherst, MA: HRD Press, 2010.

Sprinthall, R. C. *Basic Statistical Analysis.* Boston, MA: Allyn and Bacon, 2011.

Straus, E. *Phenomenology: Pure and Applied.* Pittsburg, PA: Duquesne University Press, 1964.

Truax, C. B. and Carkhuff, R. R. *Toward Effective Counseling and Therapy.* Chicago, IL: Aldine, 1967.

Watson, J. *Behaviorism.* Chicago, IL: University of Chicago Press, 1930.

Carkhuff Body of Work

Robert R. Carkhuff, Ph.D.

The Generativity Solution:
Science in the Service of Humankind

An Annotated Body of Work

CONTENTS

1. Introduction and Overview

Carkhuff was 15 years out of graduate studies in psychology when the world of science first took note of his scientific contributions in the study of helping and human relations.

Today, Carkhuff is the most powerful force in the history of science representing the science and technology of **human generativity** and formulating the **Human Sciences.**

With a profound commitment to establishing a true science of human behavior, Carkhuff has been the path-finding leader for the past 50 years as his body of work will testify.

In 1978, in terms of frequency of citations in psychology, Carkhuff ranked among the leading contributors (see Table 1). Two volumes of *Helping and Human Relations* and one volume on *Counseling and Psychotherapy* led the listings.

Table 1.
Most-Cited Books in Clinical Psychology [1]

Bandura, A. *Principles of Behavior Modification,* 1969.

Carkhuff, R. R. *Helping and Human Relations, Volumes I and II,* 1969.

Kelly, G. A. *The Psychology of Personal Constructs,* 1955.

Truax, C. B. and Carkhuff, R. R. *Toward Effective Counseling and Psychotherapy,* 1967.

Fenichel, O. *The Psychoanalytic Theory of Neurosis,* 1945

Freud, S. *Zur Geschichte der Psychoanalytischen Bewegung* (On the History of the Psychoanalytic Movement), 1914.

2. Helping and Human Relations

Carkhuff led the revolution of the helping professions from theoretical to operational treatment in the late 1960s. He and his associates defined the effective ingredients of helping in operational terms:

- *Toward Effective Counseling and Psychotherapy.* Chicago, IL: Aldine, 1967 (with C. B. Truax)

- *Sources of Gain in Counseling and Psychotherapy.* New York: Holt, Rinehart & Winston, 1967 (with B. G. Berenson)

- *Beyond Counseling and Therapy.* New York: Holt, Rinehart & Winston, 1967

Basically, the helping dimensions, such as empathy and respect, facilitated helpee exploration leading to indices of therapeutic personality change. The book with Truax (Aldine, 1967) was listed among the most-cited books in clinical psychology. [2]

Carkhuff extended these core findings to all helping and human relations. Simultaneously, he developed training programs for learning these operational skills in the practice of helping:

- *Helping and Human Relations. Volume I: Selection and Training.* New York: Holt, Rinehart & Winston, 1969

- *Helping and Human Relations. Volume II: Practice and Research.* New York: Holt, Rinehart & Winston, 1969

- *The Art of Helping.* Amherst, MA: HRD Press, 1971

Operationally, the helper's skills were refined to emphasize responding, personalizing, and initiating in order to facilitate the helpee's process, which was expanded to incorporate exploring, understanding, and acting, all of which led to therapeutic change. Now in its 9th edition, *The Art of Helping* has sold more than one million copies. In turn, the two volumes of *Helping and Human Relations* were among the most-cited books in social sciences. [3]

Along with his three books, Carkhuff was, himself, identified among the most-referenced social scientists by *The Institute of Scientific Information.* [4] [5] [6] Altogether, these works culminated the operationalization of previously theoretical processes.

3. Educational and Community Applications

In the 1970s and 1980s, Carkhuff transferred his findings to human and community development in the private as well as public sectors:

- *The Development of Human Resources: Education, Society and Social Action.* New York: Holt, Rinehart & Winston, 1971

- *Toward Actualizing Human Potential.* Amherst, MA: HRD Press, 1981

- *The Exemplar: The Exemplary Performer in the Age of Productivity.* Amherst, MA: HRD Press, 1983

- *Sources of Human Productivity.* Amherst, MA: HRD Press, 1984

Operationally, Carkhuff defined the productive human performer in the productive organizational system that, in turn, is defined in the productive community development system.

John T. Kelly, Director Emeritus, Advanced Systems Design, IBM, Inc., offered this evaluation of Carkhuff's work:

> **Carkhuff offers us a vision of the future. It is a vision of a great Age of Productivity, an age in which the human products and services are effectively increased so that all people can share. It is a vision of an age in which the resource inputs, natural and otherwise, are effectively invested so that no people are deprived of their birthrights.**
>
> (Kelly, Foreword, *Sources of Human Productivity,*
> 1984, p. xii)

During this period, Carkhuff and his associates launched a series of training products in teaching, training, and instructional systems design:

- *The Skills of Teaching Series, Volumes I–VI.* Amherst, MA: HRD Press, 1977–1981

- *Instructional Systems Design, Volumes I and II.* Amherst, MA: HRD Press, 1984

- *Training Delivery Skills, Volumes I and II.* Amherst, MA: HRD Press, 1984

Systematically, these works broke teaching and training down into four skill sets: interpersonal skills, content development skills, lesson planning skills, teaching delivery, and classroom management skills. The educational initiatives culminated in the issue of the journal, *Education,* dedicated to Carkhuff. [7]

4. Human and Organizational Processing

In the late 1980s, Carkhuff created systematic skills for human processing or thinking with applications for the development of **human and information capital:**

- *Interpersonal Skills and Human Productivity.* Amherst, MA: HRD Press, 1984
- *Human Processing and Human Productivity.* Amherst, MA: HRD Press, 1986
- *The Age of the New Capitalism.* Amherst, MA: HRD Press, 1988
- *Empowering: The Creative Leader in the Age of the New Capitalism.* Amherst, MA: HRD Press, 1989

Carkhuff's mentor, B. R. Bugelski, a protégé of Clark Hull, one of the founders of American Psychology, commented as follows:

> **This work rationalizes all of our efforts in learning theory and promises the culmination of psychology in a science of processing.**
>
> (Bugelski, Review, *Human Processing and Human Productivity,* 1984)

Carkhuff summarizes the research of the effects of individual, interpersonal, and organizational processing systems in hundreds of studies of more than 150,000 recipients. Basically, Carkhuff defined Human Capital Development or HCD as generative thinking. Carkhuff's work was reviewed by the distinguished social scientist, C. H. Patterson, University of Illinois:

> **This revolution has an important social significance also. I have stated elsewhere that the extent to which a society and its institutions, including its economic systems, facilitate the development of self-actualizing persons constitutes the criterion for the goodness of that society. To the extent that our society incorporates Carkhuff's system, it will become a better society for all its members.**
>
> (Patterson, Foreword, *Interpersonal Skills and Human Productivity,* 1983, p. 5)

5. Probabilities and Possibilities Sciences

In the 1990s and 2000s, Carkhuff, Berenson, and associates introduced the *science of possibilities* to drive the historical *science of probabilities:*

- *The New Science of Possibilities, Volume I. The Processing Science.* Amherst, MA: HRD Press, 2000

- *The New Science of Possibilities, Volume II. The Processing Technologies.* Amherst, MA: HRD Press, 2000

- *Human Possibilities.* Amherst, MA: HRD Press, 2000

- *The Possibilities Leader.* Amherst, MA: HRD Press, 2000

- *The Possibilities Organization.* Amherst, MA: HRD Press, 2000

- *The Possibilities Economy.* Amherst, MA: HRD Press, 2005

Dr. David N. Aspy, renowned scientist, educator, and protégé of Robert Oppenheimer, commented on Carkhuff's contribution to the advancement of civilization:

> **To support his views, Carkhuff does not simply offer up the science of science. He also presents the most exhaustive body of research and demonstration on relating and empowering ever presented in behavioral science. Moreover, he engaged in the most advanced demonstrations of phenomenal potential, including human, ever attempted...**
>
> **Together, these process-centric breakthroughs will lead us to a grand new Age of Enlightenment—The Age of Ideation, and in the process, The Science of Freedom.**
>
> (Aspy, *Window on the Universe, The Science of Freedom* 2007, p. 224)

6. New Capitalism and Freedom-Building

Eighteen months after having been introduced to Carkhuff's book, *The Age of the New Capitalism,* (1988), Pope John Paul II issued his Papal Encyclical, **The New Capitalism.** In it, the Pope conceded the prepotent power of **human generativity** and committed himself to the **free economy** as the only alternative after the failure of Communism:

Can it perhaps be said that, after the failure of Communism, capitalism is the victorious social system, and that capitalism should be the goal of countries now making efforts to rebuild their economy and society?

If by "capitalism" is meant an economic system that recognizes the fundamental and positive role of business, the market, private property, and the resulting responsibility for the measure of production, as well as free human creativity in the economic sector, then the answer is certainly in the affirmative, even though it would perhaps be more appropriate to speak of a "business economy," "market economy," or simply "free economy." (John Paul II, **Centesimus Annus,** 42.1–42.2, in Miller, *Encyclicals*)

Source: Carkhuff, R. R. *The Freedom Wars.* HRD Press, 2004

Early in the 21st century, Carkhuff and associates introduced the models and systems for freedom-building:

- *The Freedom Doctrine.* Amherst, MA: HRD Press, 2003
- *Freedom-Building.* Amherst, MA: HRD Press, 2004
- *The Freedom Wars.* Amherst, MA: HRD Press, 2004

In evaluating the contributions of Carkhuff to the advancement of civilization, John R. Cannon, Chief Executive Officer, Human Technology, Inc., summarizes Carkhuff's contributions [8] [9]:

> **Together, these works represent the contributions of The Possibilities Science to generating The Human and Ideational Sciences that define The Science of Freedom.**
>
> **In this context, there is nothing more powerful than the human brain enriched by possibilities experience... All of his books and all of his demonstrations, collectively from the earliest to the latest, are the products of this processing phenomenon, profound alternatives for individuals, organizations, communities, cultures, and nations; indeed, for The Global Village and its Marketplace. The Sources of Freedom are Possibilities!**
>
> (Cannon, Preface, *Science of Freedom,* 2007, p. xiii)

7. Generativity and the Human Sciences

Carkhuff's lifelong passion has been generativity or generative human processing. His recent work has focused on resolving the socioeconomic problems of our time through generative processing at all levels of community, culture, and economy:

- *The Generativity Solution, Building the New Economy.* Amherst, MA: HRD Press, 2009

- *The Generativity Solution, Volume III: Community Generativity.* Amherst, MA: HRD Press, 2009

- *The Generativity Solution, Volume IV: Cultural Generativity.* Amherst, MA: HRD Press, 2009

- *The Generativity Solution, Volume V: Economic Generativity.* Amherst, MA: HRD Press, 2009

Working with his colleagues, his body of work has culminated in a series introducing the **human sciences:**

- *Human Generativity: An Introduction to the Human Sciences.* Amherst, MA: HRD Press, 2013

- *The Human Sciences, Volume I: Probabilities, Possibilities, and Generativity Sciences.* Amherst, MA: HRD Press, 2013

- *The Human Sciences, Volume II: Probabilities, Possibilities, and Generativity Technologies.* Amherst, MA: HRD Press, 2013

- *The Human Sciences, Volume III: Carkhuff and Human Generativity.* Amherst, MA: HRD Press, 2013 (by B. G. Berenson)

Barry Cohen, Executive Vice President, Parametric Technology Corporation, summarizes Carkhuff's contributions as follows:

> **His body of research has constituted the foundation for revolutions in all areas of human endeavor: human, information, and organizational resource development: government, corporate and community development; cultural, national, and now global economic growth. In short, he has changed the world by making social science a "true science."**

> (Cohen, Foreword, *Generativity Solution,* 2009, p. ix)

Hernan Oyarzabal, Executive Director Emeritus, International Monetary Fund, summarizes Carkhuff's theory of the prepotency of generativity as follows:

> **It is indeed the "Generativity of Human Brainpower" and not the "Economic Theory of Stasis" that hold the interdependent, enlightened, and entrepreneurial keys to our Prosperous, Participatory, and Peaceful future.**

> (Oyarzabal, An Open Letter on the Economy, *Generativity Solution,* 2009, p. xiii)

8. Summary and Transition

In summary, Carkhuff's body of work has differentiated him from all others in the history of science. Primary among his contributions has been the operationalization and application of **possibilities science and generativity.**

Carkhuff's greatest contributions may lie ahead—the transfers of possibilities science to his current list of generativity projects in a troubled world. It is left to his lifelong colleague, Bernard G. Berenson, an Einsteinian scholar, to place Carkhuff's work in historical perspective:

> **Carkhuff's contributions to universal processing, alone, qualify him for leadership among the greatest scientists of history. His "nesting, encoding, and rotating of processing systems" are the core processes of Nature's Generativity. In creating The Human Sciences, Carkhuff belongs in the Pantheon of Science along with the works of DaVinci, Newton, and Einstein.**
>
> (Berenson, B. G. We Can Be Beams of Light.
> Foreword, *Human Sciences, Volume II*
> Amherst, MA: HRD Press, 2013)

In this context, Carkhuff is dedicated to the values of Albert Einstein, his exemplar:

> **It is a very high goal: free and responsible development of the individual, so that he may place his powers freely and gladly in the service of all mankind.**
>
> (Einstein, *An Ideal of Service to Our Fellow Man,*
> 1950, p. 59)

Where Carkhuff joined the list of established contributors to psychology in 1978, he now stands alone as the dominant force defining the **human, information, and organizational sciences** in the 21st century science and technology.

Along with his associates, he has architected and implemented new platforms for the following:

- **Generative human sciences**
- **Human-centric information sciences**
- **Human-centric organizational sciences**
- **Human-centric economic sciences**
- **Generative civilizations**

In so doing, Carkhuff and his associates have outperformed the entire body of the social sciences and provided leadership for the declining body of the physical sciences.

His colleague, Dr. Tom Kakovitch, a physical scientist, has the final word on generativity:

Nature grows with generativity. That is why no one holds a monopoly over intelligence.

(Kakovitch, *Collegium*, 2012)

[1]   Garfield, E. *The 100 Books Most Cited by Social Scientists.* Number 37, Institute for Scientific Information.

[2]   Garfield, E. *The 100 Books Most Cited by Social Scientists.* Number 37, Institute for Scientific Information.

[3]   Garfield, E. *The 100 Books Most Cited by Social Scientists.* Number 37, Institute for Scientific Information.

[4]   Garfield, E. *The 100 Books Most Cited by Social Scientists.* Number 45, Institute for Scientific Information.

[5]   Endler, Rushtore, and Rogdeger. *Productivity and Scholarly Impact.* American Psychologist, Vol. 33, No. 12, 1062–1082.

[6]   Heesacker, Heppner, and Rogers. "Classics and Emerging Classics." *Journal of Counseling Psychology,* Vol. 29, No. 4, 400–406.

[7]   Carkhuff, R. R. "Leader in Human Resource Development." *Education,* Vol. 106, No. 3.

[8]   Berenson, B. G. and Cannon, J. R. *The Science of Freedom.* McLean, VA: American Noble Prize, 2007.

[9]   Carkhuff, R. R. *The Age of Ideation.* McLean, VA: American Noble Prize, 2007.

www.carkhuff.com

www.mcleanproject.com

www.carkhuffgenerativitylibrary.com